Knowledge-Driven Profit Improvement

Implementing Assessment Feedback Using PDKAction Theory

Knowledge-Driven Profit Improvement

Implementing Assessment Feedback Using PDKAction Theory

Monte Lee Matthews

CRC Press
Taylor & Francis Group
Boca Raton London New York

CRC Press is an imprint of the
Taylor & Francis Group, an **informa** business

First published 2000 by St. Lucie Press

Published 2018 by CRC Press
Taylor & Francis Group
6000 Broken Sound Parkway NW, Suite 300
Boca Raton, FL 33487-2742

© 2000 by Taylor & Francis Group, LLC
CRC Press is an imprint of Taylor & Francis Group, an Informa business

No claim to original U.S. Government works

ISBN 13: 978-1-57444-229-8 (pbk)
ISBN 13: 978-1-138-46450-6 (hbk)

**Visit the Taylor & Francis Web site at
http://www.taylorandfrancis.com**

**and the CRC Press Web site at
http://www.crcpress.com**

Library of Congress Cataloging-in-Publication Data

Matthews, Monte Lee.
    Knowledge–driven profit improvement : implementing assessment
feedback using PDKAction theory / by Monte Lee Matthews.
        p. cm.
    Includes bibliographical references and index.
    ISBN 1-574-44229-5 (alk. paper)
    1. Management. 2. Knowledge management. 3. Management
information systems. I. Title.
    HD31.M3378 1999   99-22811
    658.4′038—dc21   CIP

Library of Congress Card Number   99-22811

PREFACE

This book grew out of an article I published in an international journal a few years ago. The article, in turn, was a result of lessons learned from the combined experiences of writing several quality award applications, of being a lead examiner for different state and national quality award programs, of consulting with and assisting different businesses in interpreting and using their information effectively, of listening to and reading works of Dr. W. Edwards Deming, and of being a leader of an organization responsible for running the organization smoothly and making a profit. The Plan-Do-Knowledge-Act (PDKA) theory that grew out of these experiences has been enhanced and has proven to be a highly effective tool that has been used by several organizations to improve the bottom-line of their businesses and to apply for and win prestigious quality awards.

While this book presents a radically different, yet logical way of thinking about organizational knowledge and competition, the heart of its message and its appeal is discipline, integration, and focus. I believe that every company really does have the opportunity to control and mold its own destiny, because every company does have untapped knowledge that can be tapped to steer the company toward world-class results. I believe it is possible to move beyond informational chaos to create a focused and enticing new opportunity horizon. I further believe it is possible to regenerate a clear mission with clear actions based on sound strategic knowledge, which will eliminate the crisis of chaos and misinformation.

This book presents a method for companies to take information from feedback reports and other survey, audit, and assessment sources and implement areas of opportunity into their core businesses. The strategy expands the Plan-Do-Check-Act (PDCA) logic introduced by Dr. Deming into a Plan-Do-Knowledge-Act (PDKA) process.

In Chapter 1, I explain how companies can take feedback information and convert it to usable knowledge. The knowledge can then provide the basis for implementing "assessment opportunities." By expanding the logic of the Plan-Do-Check-Act (PDCA) model into a Plan-Do-Knowledge-Act (PDKA) model, organizations can create knowledge and implement assessment opportunities.

In Chapter 2 , I explain the difference between information and knowledge. The explanation includes Dr. Deming's perspective of knowledge and begins to detail how to turn organizational information into knowledge.

In Chapter 3, I explain how results can be gained from knowledge, how indicators can be used, and how information and knowledge should be factored into the business planning cycle and tracked. I also discuss types of results, such as cultural, financial, and marketing. I conclude that most companies are acting on their improvement opportunities, but they are acting in a limited perspective; that is, they are not prioritizing or integrating the knowledge they gain from combining information and theory. Thus they will obtain positive results only by accident not by design.

In Chapter 4, I walk readers through flowcharts detailing the steps of the PDKA process. I introduce tools for organizing disparate information (the feedback summary matrix), prioritizing opportunities (the problem selection matrix), and assigning tasks and tracking their progress (the task assignment record).

In Chapter 5, I relate the principles discussed in the first four chapters to their application in the case studies that follow. I explain why I selected these case studies and what the reader can gain from them. There are many lessons to be learned—directly as well as indirectly. As a further illustration of the relationship between theory and practice, I provide matrixes showing the variety of companies represented in the case studies and which of the 12 PDKA principles they have incorporated into their own processes.

In the remaining chapters, I present a series of case studies that illustrate the principles I discussed in the earlier chapters.

I am hopeful, but not naïve, about the challenges that face any company trying to integrate information and focus on priorities. Significant challenges face any organization intent on becoming world-class by managing knowledge effectively. The first challenge, how to make use of your information by integrating it, arises because companies have many pieces of information tucked away throughout the organization, often in sub-organizations. Finding this information and bringing it together is a task in itself. The second challenge, made altogether more pressing if the first challenge is accomplished, is how to organize all of the different forms of information into a manageable framework, a framework that brings order out of the informational chaos. The third challenge is to focus equal attention on both your strengths and your weaknesses. Too often companies are focused only on improving what doesn't work well. While focusing on weaknesses is certainly important, if you neglect to build upon what is working well your strengths may soon become nothing more than mediocrities. I challenge you to build upon your strengths and make them world-class, while improving your weak spots. The fourth challenge requires the discipline to develop a decision-making criteria, a criteria based on the key company drivers, and a clear process by which to properly prioritize both your strengths and your weaknesses. You must have a crystal-clear understanding of what opportunities to focus on first, instead of trying to do everything at once. It is critical to be disciplined as you determine the top-priority items for your company to focus on and improve.

My hope is that the theory and principles presented in the first five chapters and the case studies that reinforce them will help other companies along the road to becoming world-class organizations.

The following are perspectives about the PDKA Model, from industry experts who help mold world-class organizations.

Testimonials

"The PDCA (plan, do, check, act) cycle is the fundamental concept that needs to be understood by anyone undertaking to improve the way they manage their business. It is the essence of TQM.

PDCA is an acronym standing for the scientific method and its two key components: (a) alternation between the level of thought or theory and experience or data, and (b) quest for ever better explanation of what works and why. An example is Plan what needs improving and how to go about running an experiment (level of thought or theory), Do run the experiment (level of experience or data), Check (analyze) what the results of the experiment mean (level of thought or theory), and then Act appropriately (level of experience or data) to check that way the business operates; and repeat whenever it appears a better solution is possible and worth the effort. The PDCA cycle takes business improvement out of the realm of opinion, power dominance, philosophy, and the quest for truth or idealism and puts it in the world of pragmatic performance improvement.

In his PDKA paraphrase of PDCA Lee Matthews quite properly emphasizes the importance of maximizing the understanding one gets from how the current process operates. It is in the C (or K) step that experimental data is matched to a theoretical explanation, and it is in this step that the basis is built for actually taking action. In *Knowlege Driven Profit Improvement*, Matthews also applies the PDKA cycle to his own experience and observation, turning the wisdom he has gained into an explicit theory (step-by-step) process that others can apply to their own businesses. *Knowledge Driven Profit Improvement* is a welcome addition to the improvement toolkit of business practitioners."

—**David Walden**
Executive Director
Center for Quality of Management
(retired Senior Vice President, Bolt Beranek and Newman Inc.;
Senior Lecturer, MIT Sloan School of Business; co-author, *A New*
American TQM*; retired Executive Director, CQM; editor, *Journal of
***the Center for Quality of Management*)**

"In the short period of years since Dr. Edwards Deming promoted the Plan-Do-Check-Act concept or the 'Deming wheel' it has become one of the most widely used models of continuous improvement throughout the world. It has changed the approach to Quality Management from a narrowly focused, inspection-based activity to multifaceted prevention-based discipline. Based on the powerful logic of assessment and continuous improvement this book develops the concept of knowledge as a corporate asset for achieving business excellence. To one degree or another, achieving competitive advantage depends on the organization's ability to effectively manage the information within it and to turn that information into a tree of knowledge upon which to tackle the areas for improvement and build on the organization's strengths. Since accumulated knowledge must lead to action and the creation of a new and positive perspective, the process by which knowledge is acquired and developed is of critical importance to the organization's survival. The Plan-Do-Knowledge-Act model creates a new paradigm that is effective and efficient in building knowledge and adding value to both company and individual. It does not take an enormous amount of effort, nor is it excessively complex as it builds on the PDCA model. What it does require, however, is the willingness to make decisions that will enable people to constantly add value.

With this in mind the PDKA process effectively deals with answering three basic questions:

How can we develop meaningful knowledge, rather than flooding the organization with lots of information?

How can we turn that knowledge into a corporate asset while adding value to the markets in which we operate?

How can we create a knowledge based support system in which everyone can contribute to the success of the organization?

It is these three questions that are addressed by this new approach to value-based knowledge management, without our building up knowledge for knowledge's sake. It provides the guidance, discipline and focus to move your organization toward its true competitive potential."

—Bob Barbour
Director and Chief Executive
Northern Ireland Quality Center

"The President's Quality Award Program publishes Criteria against which Federal Government organizations can conduct self assessments. These same Criteria are used to evaluate Award Program applicants. When organizations apply to the Program, they gain valuable feedback identifying their performance strengths and improvement opportunities. And by evaluating gaps between present performance and future performance goals, Program winners used the Knowledge to Plan, Do and Act model. The results were impressive! Winning organizations set and met higher performance levels and were recognized as Government models of organization excellence. Continuous improvement is a journey. But it's well worth the trip!"

—Barbara Smith
Manager
The President's Quality Award Program
United States Office of Personnel Management

"As challenging as implementing an organization-wide improvement plan often appears to private sector companies, the concept can be even more mysterious and daunting to nonprofit organizations. Segmented funding streams, diverse volunteer boards and lack of clear, measurable expectations have often contributed to nonprofit managers reaching for short-term fixes: reorganizing or throwing money at successful programs while fiscally starving struggling ones. For nonprofits who have taken a bold new direction in assessing the whole organization against a set of nationally-recognized criteria, Matthews' knowledge-based improvement plan offers a proactive, common sense approach to leverage the new insights gleaned from the assessment itself in concert with actionable data from information the organization routinely gathers.

So rather than being just the latest 'magic bullet' to appear on management's radar, the Matthews' approach combines your organization's best intelligence — new and old — and steers you toward prioritizing the information and then developing workable improvement plans on those with the greatest overall payoff. Equally important, this approach leaves a clear audit trail of the steps leading to the selection of key improvements, thus providing the leader with powerful support to motivate employees, board members and contributors. This approach to improvement planning ensures nonprofits that the sweat equity invested in the assessment process will be more than recouped with critical, lasting improvements."

—William C. Phillips
Former Senior Director, Quality
United Way of America

"In today's massive hunt to gather information which tells us how our business is doing we often get caught up in the process of assessment and feedback. We all too often lose sight of the fact that the only way for our organizations to improve is to deploy feedback into actions — appropriate actions which move our business into the competitive future. It is vital that we use our strategic knowledge derived from integrating our organizational information into a disciplined approach for strategic change.

With exposure to hundreds of organizations as director of the Tennessee Quality program, I can say that the difference between mediocre and excellent organizations is the ability to analyze and integrate knowledge into action. Understanding and acting on the 'vital few' provide the competitive edge."

—Marie B. Williams
Executive Director
Tennessee Quality Award

THE AUTHOR

Monte Lee Matthews is lead examiner for both state and national awards programs. He has had oversight responsibility for writing organizational quality award applications and implementing feedback into the organizations, and he has consulted with federal organizations, private companies, and not-for-profit industries in how best to implement their feedback in order to actually achieve improved results.

Mr. Matthews has been very active in the industry, serving as a member of the Board of Examiners for the Tennessee Quality Award, the Board of Examiners for the United Way of America's Excellence in Service Quality Awards, and the Board of Examiners for the President of the United States Quality Award.

Mr. Matthews has a B.A. in Psychology from the University of North Carolina at Chapel Hill, a B.A. in Architecture from the University of North Carolina at Charlotte, and a Master's in Architecture with a Minor in Management from The School of Design at North Carolina State University. He is also an alumnus of the Executive Programs at Columbia University, The Amos Tuck School at Dartmouth College, The Wharton School at The University of Pennsylvania, and the J. L. Kellogg Graduate School of Management at Northwestern University. He is a Board Certified Licensed Architect.

Mr. Matthews is currently Senior Strategist for the Electric System Operations of the Tennessee Valley Authority (TVA). He is responsible for business and strategic planning, transforming the business and monitoring performance indicators.

Mr. Matthews has been intimately involved with strategic planning and short-term business throughout his career. He was Manager of Quality Programs for the Fossil and Hydro Power organization within the largest U.S. electric utility. During his tenure, the Power organization was awarded

Finalist for the President of the United States Quality Award and was a winner of the Tennessee Quality Award.

Mr. Matthews has also been a Quality Manager for the Chief Operating Officer of a $5.6 billion company, where he supported the strategic and tactical planning function of the Executive Council, recommended and monitored the development of business improvement strategies, led the business alignment initiative, and provided critical input to the Board of Directors.

Mr. Matthews has served as Chief Architect and Senior Manager of a Diversified Engineering and Technical Services organization where he was responsible for leading and overseeing the financial management, strategic marketing, and strategic initiatives leading to business growth. He managed domestic and international contractual relationships, as well as overseeing engineers, quality assurance specialists, architects, and project managers.

Mr. Matthews has presented at national and international conferences and to executive audiences at their invitation, and has lectured at many universities and colleges. Mr. Matthews was selected to Who's Who in the World, and chosen as one of the "Outstanding Young Men of America." He is a volunteer for Habitat for Humanity and the Appalachian Service Project, and has a passion for mission work.

ACKNOWLEDGMENTS

I wish to thank the following individuals for their contributions to this book.

Judy Pearson, my editor and sounding board for ideas, is a co-worker I met during my first attempt at coordinating the massive effort of applying for a quality award. Since then we have worked together on other successful writing projects. It is a rare gift to find an editor who knows what you really want to say and helps you say it. This book would not have been possible without the dedicated efforts of Judy Pearson, who spent countless hours in assisting me in refining the book's materials.

Many thanks to Bob Barbour, who was the key in obtaining case studies from Europe as well as coordinating with the European Commission and The European Foundation for Quality Management. Thanks to Barbara Smith for her support and guidance. Many thanks to all of the case study participants for their willingness to participate in this book, and to the key individuals who actually prepared the case studies for their companies.

Thanks to my family: to my mother and father for instilling the drive in me to write this book, and to my wife Debbie and our children Christopher and Kellie for allowing Daddy to spend time — normally dedicated to them — on writing this book. Above all, thanks to God for blessing me with His many gifts.

CONTENTS

1

THE LOGIC AND PHILOSOPHY BEHIND THE PDKA MODEL

INTRODUCTION

This chapter explains the logic and philosophy behind the Plan-Do-Knowledge-Act (PDKA) logic and process, including perspectives from leading experts, such as Dr. W. Edwards Deming. The logic will explain how companies can take feedback information and convert it to usable knowledge. The knowledge will then provide the basis for implementing "assessment opportunities." This chapter will also explain the Plan-Do-Check-Act (PDCA) model and how the PDKA model was derived from it. By expanding the logic of the PDCA model into a Plan-Do-*Knowledge*-Act model, organizations can create knowledge and implement assessment opportunities.

THE PLAN-DO-CHECK-ACT LOGIC

Following the PDCA model to improve business can appear to be a simple task, but the model's simplicity is deceptive. The fact is that most businesses do not have the discipline or the proper business knowledge to implement their improvement opportunities, and even if they do, good results are often achieved more by accident than by design.

The Plan-Do-Check-Act model is based on the concept that "the cornerstone of rapid learning is Plan-Do-Check-Act (PDCA)."[1] According to Brian Joiner, "the basic notion of PDCA is so simple that when I first heard it I felt I understood it in five minutes. Now, more than a decade later, I think I might understand it some day."[1]

1

What Does "Plan-Do-Check-Act" Mean?

P: Plan what you're going to do and how you will know if it works.
D: Do; carry out the plan.
C: Check; evaluate the outcome, learn from the results.
A: Take action.[1]

The Plan-Do-Check-Act (PDCA) logic is based on the concept that in order to achieve improvement to an organization, the organization needs to plan for it, implement it, check the results, and take action for further improvement.

The Wheel Concept

This Plan-Do-Check-Act concept was promoted by Dr. W. Edwards Deming and has come to be known as the *Deming Wheel*. Deming based this concept on earlier work by Dr. Walter Shewhart. The PDCA concept is also sometimes referred to as the Shewhart Cycle, with the steps being Plan-Do-Study-Act. "The PDCA wheel is a simple, flexible model that has been adapted to a broad range of situations."[1]

Turning the Plan-Do-Check-Act Wheel

The Plan-Do-Check-Act logic thrives on new ideas—ideas about how to make things work better. These ideas are often the result of multiple forms of information intersecting for the first time that cause the light bulb to go on and new organizational knowledge to be determined. Without new ideas surfacing in companies, there will be no significant improvement. New ideas come when information is integrated to gain true knowledge.

The PDCA wheel can turn weekly, monthly, or yearly. Here are some examples.

Within a week, a custodian solved the problem of people throwing trash in recycling bins intended for aluminum cans. He made a hole the size of an aluminum can in some lids and wrote "aluminum cans only" on these containers. He found little or no trash in these bins, so he implemented the solution using all of the containers. "With one turn of the PDCA wheel he made a plan, carried out a small-scale test, and checked the results. With another turn he spread the solution to all recycling bins."[1]

Yearly, a management team examines its planning process. Team members begin by examining the previous years' activities and determining how accurate their plan had been. They look at what had worked well as well as what had gone wrong. In one such evaluation they discovered that they had taken on too many projects and had not allotted enough time for them. Upon learning this, they improved their process to avoid the same problems. Each year the company improves its processes by planning, doing, checking and acting.[1]

PDCA is at the core of continuous improvement. It is based on the fact that no system or process is perfect and that all business systems must be checked, improvement areas identified and prioritized, and solutions deployed, starting with the most critical improvement opportunities.

> PDCA is the scientific approach applied to the workplace. . . .
> [It] is the essence of managerial work: making sure the job gets done today and developing better ways to do it tomorrow. . . .
> How often, how fast, and how effectively we rotate the PDCA cycle determines how well we get the job done and how fast we improve.[1]

The cycle begins with an assessment of the current situation. Information is gathered and a plan for improvement is developed (Plan). When the feedback is in and the plan is finalized, improvements are implemented (Do). Actual performance is checked to see if implemented improvements have actually resolved the problems (Check). If the problems have been solved, then the improvement initiative and the process associated with it is standardized (Act). After this cycle, the organization plans for further improvement, and repeats the cycle. This is called turning the Deming Wheel. If the plan was not successful, then it is assessed to determine why it was not successful, and a new plan is developed and the cycle begins again. In a business' daily operations, everything should be in one of these steps of the PDCA cycle—planning, doing, checking, or acting for further improvement.

Keys To Using PDCA Effectively

Using PDCA logic to review a business system, then advancing that logic by integrating all of the findings together, can provide an organization with invaluable knowledge related to the comprehensive performance level of its operations.

According to Joiner, there are several keys to using PDCA effectively:[1]

■ Adapt to the situation.
■ Challenge yourself to find ways to do small-scale tests quickly.

For example, Joiner suggests starting with a paper and pencil analysis and narrowing the field to "one or two alternatives." After discussing the alternatives with those who would be affected by the change, perform a walkdown, refine, and implement the results on a trial basis.

■ Decide ahead of time how you will assess progress.[1]

Failure to know what results you are looking for and when you have attained them is wasteful. Take careful notes, especially concerning that which does not go according to plan. Then, as Joiner stresses, be sure to study the results. He adds that "to be effective, the entire PDCA cycle must become the basic mindset of every employee, every department, every function."[1] Too often the elements of the PDCA cycle, Planning, Doing, Checking, and Acting, are carried out by different departments and managers. The more fragmented the PDCA functions are, the more impossible it is to turn the PDCA wheel.

The PDCA logic, coupled with using an established criteria as a framework for organizing all of the assessment feedback, can enable an organization to step back periodically and assess how the complex inter-actions within the organization are working or not working. From the assessment findings, an organization can establish both strategic and operational plans critical to its business success and work toward the integration of all its activities consistent with world-class objectives. The assessment process also allows members of the organization a chance to grow and learn about their business operations and how their efforts can be focused on systematic improvement. The problem is that most businesses have neither the discipline nor the method necessary to implement their improvement opportunities.

The Importance of the Check Step

One common approach companies have used to make improvements, according to Mark Graham Brown, is to take each area of opportunity and develop action plans for all of the opportunities.[2] Committees and task forces are formed and hundreds of people are involved in trying to improve the organization's performance, often in more than a hundred different areas. They spend a year and thousands of dollars in labor working on the various improvement projects. They do the assessment again the following year, and, to everyone's surprise, the overall score does not improve much. The reason for the failure is that such an approach is too diluted and too uncoordinated. It lacks tie-in to the company's strategic business plan. Teams end up stepping on each other's toes, perhaps improving performance in one area only to make it worse in another.[2]

In other words, companies have become good at Planning and gathering information about their businesses, and they have become good at Doing assessments of their businesses; however, they typically stop the PDCA cycle once the feedback report from that assessment is in their hands. They stop short of Checking their feedback data and setting Actions that could drive the improvement of their organization's entire business system, including improving the bottom line. According to Brown, "correcting audit findings is the most critical phase in the audit process; unfortunately, it is also the phase most often done poorly."[2]

Joiner agrees that the check step is extremely important. When some executives complained that they could not get their organizations to plan, he asked them about "C," the Check step: "It turned out that they'd never given the C, the Check step of PDCA, any attention at all, so they had never learned what they were doing wrong or where they could improve."[1]

According to Joiner,

> All of the PDCA cycle is important for rapid improvement. But among the steps, "C" (Check) is the driver of rapid learning. Without it, improvement is nearly impossible. Performing a Check is something that few organizations do regularly or well. Instead, they only execute the Plan and Do portions of PDCA—with the emphasis heavily weighted toward Do! This incomplete execution of PDCA is what many people think of as "decision making." By getting conscientious about Check, by treating decisions as experiments from which we must learn, we get all the components of PDCA to fall into place.[1]

As critical as the check step is, it is even more vital when an organization has multiple types of information that are derived by checking. This information must be integrated and checked and balanced against other information to ensure that the checking is effective. This advanced level of checking and intertwining all findings leads to true organizational knowledge.

The Limitations of the PDCA Model

PDCA is a sound logic for use within each of the systems in any business; however, the logic, as it applies to the mass of information in most businesses today, needs to be modified so that all of the information is integrated for the business as a whole. In other words, rather than conducting separate PDCA cycles for each of the information feedback avenues, such as employee surveys, employee focus groups, customer surveys, customer focus groups, supplier surveys, supplier focus groups, self-assessments, quality award assessments, and business indicator reviews, the maturing of the PDCA logic as it relates to information occurs by *integrating* all information to create Knowledge. This Knowledge step, which replaces the Check step, eliminates the potential duplication of effort and allows parallel activities to be coordinated to reduce effort and time. The most important benefit of this Knowledge step, however, is that it prioritizes all of the business opportunities in a coordinated fashion, including both the weaknesses and the strengths. From this point, all future information can be integrated into this same PDKA process, ensuring continuous bottom-line improvements.

THE PLAN-DO-KNOWLEDGE-ACT LOGIC

The PDKA model that grows from the PDCA logic is made up of four simple steps:

1. PLAN for reviewing the business system's eventual improvement.
2. DO an assessment or survey that leads to obtaining feedback.
3. KNOWLEDGE is the critical piece of this model. Knowledge is the integration of information with practical application, it is the synthesis of new ideas generated from the information. In this step, the CHECK step of the PDCA model becomes the point where all organizational information is integrated to become actual usable knowledge for the business.
4. ACT to deploy solutions based on the knowledge obtained and track the results of those solutions.

A key factor in understanding the Plan-Do-Knowledge-Act logic is the ability to understand the difference between *Information* and *Knowledge*. Information comes in many forms, is often disjunctive, and often overloads companies to the point that it is rarely used effectively:

> Information, no matter how complete and speedy, is not knowledge. Knowledge has temporal spread. Knowledge comes from theory. Without theory, there is no way to use the information that comes to us on the instant.[3]

Many organizations have not yet figured out how to effectively take the information from their variety of information sources—Baldrige-style assessments, employee surveys, customer feedback, indicator results, or any others—and actually use that information for improvement planning.

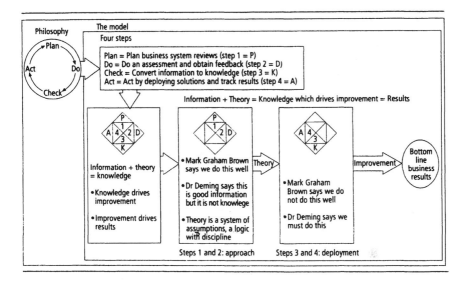

Figure 1.1   PDKA Philosophy

Many companies have not figured out how to put teeth into their feedback and ensure that areas for improvement become integrated into the overall business plan and system logic of the organization.

Businesses need to couple the information that they gather from audit findings with theory. When they combine information with theory, they create knowledge—true knowledge that will drive disciplined improvement, which will in turn drive results.

> PDKA Logic : Information + Theory = Knowledge
>
> Knowledge → Improvements → Results

Figure 1.1 shows the philosophy of the PDKA model.

Steps 1 and 2 of this model (Plan and Do) lead a company to information. This information is indeed important, because without information, the company cannot obtain knowledge, but this information by itself is not knowledge, and the information by itself will not lead an organization to success. The information usually exists in several different subgroups within the organization, for example, customer surveys in the marketing department, employee surveys in human resources, key business indicators in business operations, and national award feedback assessments in quality resources.

Steps 3 and 4 (Knowledge and Act) require sound logic and discipline. To effectively deploy the right solutions for the business, this disparate information must be combined using a disciplined theory or organizing framework, such as the seven Baldrige criteria (see Chapter 4) to create new knowledge. Both weaknesses and strengths must be prioritized and integrated into business processes. Finally, solutions to these problems must be established in the business to achieve world-class results.

CONCLUSION

In summary, the PDKA model shows businesses how to integrate all information and convert all identified opportunities into knowledge, prioritize the knowledge before proceeding to develop action plans, then act to deploy the solutions to these opportunities into the business. With this approach, the organization selects only the most important areas of opportunities for the next year. These areas of opportunity are linked to a problem-solving process and to a business planning process.

Businesses that have already implemented the PDKA model have been very successful in the competitive arena. Many are winners of prestigious business awards such as the Baldrige award. The success of such companies was demonstrated in Ray Helton's "The Baldie Play." He hypothetically invested a thousand dollars in each publicly owned Baldrige Award winner and saw an increase in his investment of almost 100%.[4] As a result of systematic improvements based on implementing solutions to the most critical opportunity areas, world-class companies have made significant improvements to their bottom lines. With some planning and hard work, along with proper business knowledge, *your* company can be world-class.

2

TURNING KNOWLEDGE
INTO A CORPORATE ASSET

INTRODUCTION

As explained in Chapter 1, the cornerstone of business improvement has been the Plan-Do-Check-Act (PDCA) logic. This logic is based on the concept that in order to achieve improvement, an organization must not only plan for improvement, it must implement actions for further improvement. The sad fact is that most businesses do not have the discipline or the proper business knowledge to implement their improvement opportunities, or if they do implement plans, their results are less than expected. There are reasons for poor or uneven results. Although many companies have become good at *plan*ning and gathering information about their businesses, and even good at *do*ing assessments of their businesses, they typically stop the PDCA cycle once the feedback report from that assessment is in their hands. They stop short of *check*ing the feedback. Even more importantly, they fail to analyze their data and combine it with an organizing theory to obtain usable knowledge. Thus, they never set meaningful *act*ions that could drive the improvement of their organization's entire business system, including improving the bottom line. This is the value of *knowledge*.

To put the power of knowledge in perspective, consider Dr. W. Edwards Deming's answer to a question posed to him in a seminar:

Question: Please elaborate on your statement that profound knowledge comes from outside the system. Aren't people in the system the only ones that know what is happening and why?

Answer: The people that work in any organization know what they are doing, but they will not by themselves learn a better way. Their best efforts and hard work dig deeper the pit that they are working in. Their best efforts and hard work do not provide an outside view of the organization. The theory of knowledge teaches us that a statement, if it conveys knowledge, predicts future outcome, with risk of being wrong, and that it fits without failure observations of the past. . . . Information, no matter how complete and speedy, is not knowledge. Knowledge has temporal spread. Knowledge comes from theory. Without theory, there is no way to use the information that comes to us on the instant . . . A dictionary contains information, but not knowledge. A dictionary is useful. I use a dictionary frequently when at my desk, but the dictionary will not prepare a paragraph nor criticize it.[3]

According to Dr. Deming, an "outside view" of the organization means the ability to really understand what needs to be done differently to a business so that it will actually improve. In other words, the outside view is really a fresh view, which brings with it confidence that new order and sound logic can be applied through knowledge so that an organization can actually improve. Wherever the outside view comes from, it must be based on knowledge. Businesses need to couple the information that they gather with theory. In other words, businesses need to convert their assessment and survey feedback into knowledge. When they combine information with theory, they create knowledge—knowledge that can move their businesses into the future.

TURNING INFORMATION INTO KNOWLEDGE

The Difference between Information and Knowledge

A key factor in understanding the value of knowledge is the ability to understand the difference between *information* and *knowledge*. Information comes in many forms, is often disjunctive, and often overloads companies to the point that it is rarely used effectively. Organizations must figure out how to effectively take the information from their variety of information sources—Baldrige-style assessments, employee surveys, customer feedback, indicator results, or any others—and actually integrate that information into usable knowledge, which can then be used effectively in improvement planning. When information is combined with *theory*, knowledge is created—knowledge that can move businesses into the future.

Knowledge is the state or fact of knowing, the familiarity, awareness, or understanding gained through experience or study. Knowledge has temporal spread and incorporates theory and information over periods of time. Synonyms of knowledge are wisdom, learning, and enlightenment. Information in and of itself is not knowledge; information is the act of informing or the condition of being informed. Information and data are somewhat synonymous: data is information, especially information organized for analysis or used as a basis for a decision. Neither data, information, nor intelligence—the capacity to acquire and apply knowledge—is equivalent to knowledge.

How Intelligence Leads to Knowledge

Intelligence is the capacity to acquire and apply knowledge. It is this ability that is essential to any business that wants to become and remain competitive. A company acquires knowledge by integrating all information, by prioritizing existing strengths and weaknesses, and by determining what things to focus on to improve the company. Thus the integration of business intelligence, business ideas, and business information leads to knowledge, knowledge that will drive the company's improvements.

Intellect, although it is not knowledge, is the temporal spread that allows companies to obtain knowledge. Intellect, as explained by James Brian Quinn, "is the core resource in producing and delivering services."[5] As a company produces and delivers more and more services and products, and as the effectiveness of the processes behind the services and products becomes more and more critical to the very existence of the company, organizational intelligence becomes critical to the company's success.

Quinn makes the following points regarding the importance of intellect in the U.S. economy.[5]

- Provision of services has become by far the largest component of the U.S. economy (74% of the Gross National Product) and dominates manufacturing in virtually all "industrialized" nations.
- Through 1990, the services sector continued to grow during recessions and booms alike, although there were some signs that its rate of growth might be slowing and that it too might become vulnerable to downturns.
- In contrast, total employment in manufacturing had declined erratically but in a continuous downward secular trend over the past 15 years.
- The basic cause of this massive economic transformation is the emergence of intellect and technology—particularly in services—as highly leverageable assets.

As intellect and technology are becoming highly leverageable assets, so must organizational information be transformed into organizational knowledge in order to leverage the business against competition. Intellect, or the intelligence within business, is parallel to the knowledge gained through applying the Plan-Do-Knowledge-Act logic to an entire business system.

A KNOWLEDGE-BASED BUSINESS SYSTEM

In order for a company to be driven to actually obtain bottom-line improvements, a transformation must occur. This transformation requires knowledge based on sound organizing principles such as the Baldrige criteria, the checklist a company uses to evaluate itself before applying for a Baldrige Award. Knowledge comes from a variety of information sources, which come from many parts of a system and which should be integrated through a sound logic. The many individual parts of any business system should reinforce each other to accomplish the company's goals. Unfortunately, the business' systems and the subsequent business information often work against each other. The goal most companies are attempting to reach is to improve the business systems and thus the business' bottom line. However, a system cannot understand itself. So, the transformation of information to knowledge within any business system will require the discipline of a sound logic, the intelligence of its leaders to understand the importance of implementing the logic, and the desire of its leaders to actually drive the improvements gained from the knowledge.

This business transformation will occur when the company

- Obtains information about pluses and minuses of its business;
- Integrates its information by using a sound logic or theory by which to organize and prioritize its opportunities; and
- Implements this new organizational knowledge, resulting in bottom-line improvements.

Optimizing the System

Information comes in a variety of forms within a business, and the information must work in concert to achieve goals for the business. Unfortunately, in most businesses the information sources do not work in concert, organizational goals are not obtained, and positive bottom-line results not realized. The effective use of information depends on having a system within the business. Also, there must be an understanding of variation, which is critical in recognizing that different information sources reach different conclusions and that the bundles of information must be transformed into usable knowledge.

There are not more than five musical notes, yet the combinations of these five give rise to more melodies than can ever be heard. There are not more than five primary colors, yet in combination they produce more hues than can ever be seen. There are not more than five cardinal tastes—sour, acrid, salt, sweet, bitter—yet combinations of them yield more flavors than can ever be tasted.[6]

Likewise, the integration of information will lead to an exhausting combination of possibilities for strategic knowledge. A business system itself is an exhaustive combination of information possibilities, and a business system is a network of interdependent parts that work together to try to accomplish bottom-line results.

A system must have a focus or a goal to achieve. Without a clear focus, there is no system. The goal for most businesses is to achieve bottom-line improvements or to become increasingly more competitive. A system must also be managed effectively and be optimized. *Optimization* is a process of orchestrating the efforts of all components toward achievement of the stated goals. In other words, a sound logic and discipline must be followed. A system must not only follow the Plan-Do-Check-Act logic, it must expand that logic to use its information *knowledgeably.*

An example of a system, well optimized, is a good orchestra. The players are not there to play solos as prima donnas, each one trying to catch the ear of the listener. They are there to support each other. Individually, they need not be the best players in the country.

Thus, each of the 140 players in the Royal Philharmonic Orchestra of London is there to support the other 139 players. An orchestra is judged by listeners, not so much by illustrious players, but by the way they work together, the way that they integrate. The conductor, as manager, begets cooperation between the players, as a system, every player to support the others.[3]

So, too, an optimized business system is one that fosters cooperation among the "players." The obligation of any part of the business system is to contribute its best to the system, not to maximize its own production, profit, or sales. The whole system is optimized when all of its members allow information to become knowledge.

Understanding the Variety of Information

Organizations must grasp the importance of understanding the varieties of information they may have. Companies must acknowledge that there will always be variation in the information sources. Any two pieces of information that assess the same issue from varying perspectives will have variation in the findings. It is important for a business to understand the variations in its business and to determine what the variations mean to the results of the whole business system.

Often, the scattered information and the lack of integrated information is an indication that the business system is out of control. It is vital that the method by which businesses effectively use their information be controlled and thus predictable, so that they can rely on conclusions based on that information. Thus a challenge for businesses is to bring their varieties of information into statistical control, so that all of the information together will define future capability and thus the possibility of attaining improved future performance. In other words, the information must have a definable capability, and it must be organized by some framework that allows eventual integration and prioritization.

Two costly mistakes frequently made in attempts to improve results are to

- React to an outcome as if it came from a special cause, when actually it came from common causes of variation.
- Treat an outcome as if it came from common causes of variation, when actually it came from a special cause.[3]

The information that companies acquire and on which they typically act is most often not understood in terms of cause and effect. If all information sources are integrated to produce true knowledge using the PDKA method explained in this book, companies will know which issues are due to special causes or common causes and, in fact, which causes are the root of larger organizational problems.

Creating Strategic Knowledge

As stated previously, information by itself is not knowledge. Information *plus theory* equals knowledge, knowledge that can drive improvements and thus bottom-line results in a business. In *The New Economics for*

Industry, Government, and Education, Dr. Deming stresses the idea that knowledge is built on theory:

> Rational prediction [of future results] requires theory and builds knowledge through systematic revision and extension of theory based on comparison of prediction with observation. The barnyard rooster Chanticleer had a theory. He crowed every morning, putting forth all his energy, flapped his wings. The sun came up. The connection was clear: His crowing caused the sun to come up. There was no question about his importance. There came a snag. He forgot one morning to crow. The sun came up anyhow. Crestfallen, he saw his theory in need of revision. Without his theory, he would have had nothing to revise, nothing to learn.[3]

It is extension of application that discloses the inadequacy of a theory and the need for revision or even for a new theory. Without a logic or a theory, there is nothing to revise. Without theory or sound logic, there is no grounding by which to make changes. In other words, the experiences within our businesses have no meaning. Without theory, there are no questions to ask. Hence without theory, there is no learning. So, in order to learn, an organization must take the random information that exists and unite it by way of a sound logic or theory to create knowledge.

An example of a theory may be using Baldrige-type criteria (Chapter 4) to guide the review and assessment of a business system. Adding theory to information gathered from a logic, such as the Plan-Do-Check-Act logic, affords an organization a more comprehensive view of its operations. It offers a better view of the complex interactions within the organization combined with the outside perspective of the assessment and the Baldrige criteria. In effect it is the disciplined process applied to organizational information to convert it into strategic knowledge built into the Plan-Do-Knowledge-Act logic.

According to Deming, theory offers a window into the world, which leads to *prediction.* Without prediction, experience and examples teach nothing. Any rational plan, however simple, is prediction concerning conditions, behavior, performance of people, procedures, equipment, or materials.

Use of data or information requires prediction. Interpretation of data or information from a test, an experiment, or a survey is prediction—what will happen on application of the conclusions or recommendations that are drawn from a test or survey? This prediction will depend largely on knowledge of the subject matter. It is only in the state of statistical control that statistical theory provides, with a high degree of belief, prediction of performance in the immediate future. So, how are most companies able to take a variety of information sources, make scattered decisions on what to improve, and then show actual improvements? The answer is that if improvement actually occurs in this scenario, it is by chance, not based on a knowledgeable rationale.

Again, information is not knowledge. Companies are today able to instantly communicate large quantities of information with any part of the world. Unfortunately, speed does not help anyone understand the information or to make knowledgeable decisions to help predict the future.

Using Knowledge to Drive Improvement: The Twelve Principles

The way to drive improvement is through the use of organizational knowledge; that is, the knowledge that comes from the integration of all organizational information. Most organizations are not disciplined enough to take the various information sources, such as assessments, employee surveys, or customer feedback, and use it for improvement planning. Most companies are not making effective use of their assessments, or ensuring that areas for improvement are becoming aligned with the overall business plan and system logic of the organization.

The Plan-Do-Knowledge-Act process converts all identified feedback opportunities into knowledge, prioritizes the knowledge before developing action plans, then implements the solutions to these opportunities into the business. Based on the prioritization process, the organization selects the most important areas of opportunities. These areas are implemented using a type of problem-solving process and are aligned to a business planning process so that they can be continuously monitored until the improvement becomes standardized.

There are essentially 12 principles that businesses must use to convert information to knowledge and thus drive bottom-line improvements. And among these 12 principles are 5 that are core to the PDKA philosophy; these are shown in italics in the list below.

1. Make sure that senior executives clearly understand the organization's feedback and the need to create knowledge.
2. Form an action plan team to pull together all information.
3. *Consolidate Information*: assemble and consolidate all information sources and feedback.
4. *Use an Organizing Framework*: use quality award categories as an organizing framework.
5. *Identify Both Strengths and Weaknesses* to improve.

6. *Develop Decision-Making Criteria* by which to measure priority.
7. *Prioritize Strengths and Weaknesses*: prioritize the lists of strengths and weaknesses, and attack the strongest strengths and the weakest weaknesses.
8. Assign selected opportunity areas as tasks and select a sponsor and a team leader for each task.
9. Implement the improvements for each task by using a type of problem-solving process.
10. Track the progress of all tasks.
11. Conduct reviews periodically, and integrate the actual improvements into the business planning process.
12. Standardize the improvements.

These 12 principles are the key ingredients to the success of implementing the PDKA logic.

Unfortunately, most businesses do not implement a Plan-Do-Knowledge-Act logic, whether it relates to their business planning, their problem solving, their strategic planning, or any other part of their business. In essence, companies are willing to invest time in planning and gathering information about their businesses, but they do nothing with their critical information. They stop short of creating knowledge because they do not integrate their organizational information with theory to produce knowledge. As a result, they fail to establish actions that could drive the improvement of their organization's entire business system and improve the organization's bottom line. It is critical that businesses transform their information into knowledge and use that knowledge to drive improvement.

3

THE IMPORTANCE OF BOTTOM-LINE RESULTS

INTRODUCTION

No matter how hard companies try to convince themselves otherwise, businesses are driven by bottom-line results. A simple definition of a result is something that occurs or exists as a consequence of a particular cause. To result in something is to end in a particular way: bottom-line results state how much money has been made or lost or whether the company returned a profit for its stockholders. The word "result" is synonymous with the word "effect." The reason businesses do not achieve the results they want is because, by definition, something caused them not to achieve the results they had hoped for.

Results become more important when a company is competing against other companies. If company A's results are less than company B's results, then company B is the winner and company A is the loser. Human nature causes managers of companies to want to win. The existence of competition makes a company want to have business results that are as good as possible. In this respect, companies want to win in the game of comparing results. As expressed by the Chinese philosopher Sun Tzu, "what the ancients called a clever fighter is one who not only wins, but excels in winning with ease."[6]

In the business world, winning may be the same as a company's being the best by consistently obtaining world-class results. Many companies fail to realize that because conditions within a company change constantly, the information that defines these conditions becomes the avenue by which they can seize control of the business and obtain the world-class results they desire. As expressed by Sun Tzu,

Just as water retains no constant shape, so in warfare [business] there are no constant conditions. The five elements—water, fire, wood, metal, earth—are not always equally predominant; the four seasons make way for each other in turn. There are short days and long; the moon has its periods of waning and waxing. He who can modify his tactics in relation to his opponent [competition], and thereby succeed in winning, may be called a heaven-born captain.[6]

To be realistic, in a business setting anything being considered will need to show cost/benefit or return on investment. The benefits may be indirectly or directly tied to bottom-line dollars. If a business cannot demonstrate how an initiative is worth doing, there is little chance that the initiative will ever start. It is absolutely critical that results are gained when a company initiates an action. The Plan-Do-Knowledge-Act process will translate business information into knowledge that will move the company toward achieving improved bottom-line results.

TYPES OF BUSINESS RESULTS

Business results refers to outcomes in achieving the goals of an organization. The following factors are used in Baldrige-style assessments to evaluate results:

■ Current performance levels
■ Performance levels relative to appropriate comparisons and/or benchmarks
■ Rate, breadth, and importance of performance improvements
■ Demonstration of sustained improvement and/or sustained high-level performance

As defined in Baldrige-style assessment criteria (i.e., Malcolm Baldrige National Quality Award, the President of the United States Quality Award Program, the State of Tennessee Quality Award, or the United Way's Excellence in Service Quality Award), the business results to be examined are the organization's performance and improvement in five key areas, each of which will be discussed below: (1) customer satisfaction; (2) financial and market sector performance; (3) human resources; (4) supplier and partner performance; and (5) company-specific operational performance. Also examined are how a company's performance levels compare to those of its top competitors.

Customer-Related Results

This key indicator addresses the principal customer-related results of customer satisfaction, customer dissatisfaction, and customer satisfaction relative to competitors. The indicator calls for the use of all relevant data and information to establish the company's performance as viewed by the customer. Relevant data and information include customer satisfaction and dissatisfaction; retention, gains, and losses of customers and customer accounts; customer-perceived value based on quality and price; and competitive awards, ratings, and recognition from customers and independent organizations. Examples could be customers lost per year compared to those lost by competitors or level of customer delight in a product compared with that in a competitor's product.

Financial and Market Results

This key indicator summarizes the results of the organization's key financial and market sector performance by providing current levels and trends for all quantitative measures and/or indicators of performance. These results should always include appropriate comparative data from competitors to determine the organization's ranking in the market. Results related to financial performance include aggregate measures of financial return and/or economic value, and results related to service sector performance, including sector share, business growth, and new sectors entered. Such key financial and market measures are frequently tracked by senior organization leaders on an ongoing basis to gauge overall company performance and to determine their incentive compensation. Measures of financial performance can also include return on equity, return on investment, operating profit, pre-tax profit margin, earnings per share, profit forecast reliability, and other liquidity and financial activity measures. Marketplace performance can include market share measures of business growth, new product and geographic markets entered, and percent new product sales, as appropriate.

Human Resource Results

This key indicator summarizes the organization's human resource results, including employee well-being, satisfaction, development, and work system improvement and effectiveness. Results can include generic and business- or company-specific factors. Generic factors include safety, absenteeism, turnover, and satisfaction. Business- or company-specific factors include those commonly used in the industry or created by the company for purposes of tracking progress. Results reported might include input data, such as extent of training, but the main emphasis should be placed on measures of effectiveness. Results reported for work system performance should include those relevant to the company and might include measures of improvement in job classification, job rotation, work layout, and changes in local decision making. The indicator calls for comparative information so that results can be evaluated meaningfully against those of competitors or other relevant external measures of performance. For some measures, such as absenteeism and turnover, local or regional comparisons are also appropriate.

Supplier and Partner Results

This key indicator summarizes the organization's results of supplier and partner performance, which include identifying current levels and trends in key measures and indicators of supplier and partner performance. Also, the results should include organization cost and/or performance improvements attributed to supplier and partner performance. The focus should be on the most critical requirements from the point of view of the company: the buyer of the products and services. Data reported should reflect results by whatever means they occur—via improvements by suppliers and partners and/or through selection of better performing suppliers and partners. Measures and indicators of performance should relate to the principal factors involved in the company's purchases—quality, delivery, and price. Data reported should also reflect how suppliers and partners have contributed to the company's performance goals. Results can include cost savings; reductions in scrap, waste, or rework; and cycle time or productivity enhancements. The indicator calls for comparative information so that results reported can be meaningfully evaluated against those of competitors or of other relevant external measures of performance.

Company-Specific Results

This indicator summarizes operational performance results that are not covered by other indicators but that significantly contribute to key organization goals such as customer satisfaction, operational effectiveness, and financial performance. Key company-specific results are derived from operations and related product quality and performance; key process performance; productivity, cycle time, and other effectiveness and efficiency measures; regulatory/legal compliance; and other results supporting the organization's strategy, such as new product/service introductions.

Measures and/or indicators of product and service performance should relate to requirements that matter to the customer and to the marketplace. If the features have been properly selected, improvements in them should show a clear positive correlation with customer and marketplace improvement indicators. The correlation between product/service performance and customer indicators is a critical management tool: a device for defining and focusing on key quality requirements. In addition, the correlation might reveal emerging or changing market segments, the changing importance of requirements, or even the potential obsolescence of products and/or services.

Product and/or service performance appropriate for inclusion might be based upon internal company measurements, field performance, data collected by the company or on behalf of the company, or surveys of customers on product and service performance. Although data appropriate for inclusion are primarily based upon internal measurements and field performance, data collected by the company or other organizations through follow-up might be included for attributes that cannot be accurately assessed through direct measurement (for example, ease of use) or when variability in customer expectations makes the customer's perception the most meaningful indicator.

Measures and/or indicators of operational effectiveness could include environmental improvements reflected in emissions levels, waste stream reductions, by-product use, and recycling; cycle time, lead times, set-up times, and other responsiveness indicators; process assessment results such as customer assessment or third-party assessment (such as ISO 9000); and business-specific indicators such as innovation rates, innovation effectiveness, cost reductions through innovation, time to market, product/process yield, complete and accurate shipments, and measures of strategic goal achievement.

The company-specific results indicator encourages the use of any unique measures the company has developed to track performance in areas important to the company. It calls for comparative information so that results reported can be evaluated against those of competitors or of other relevant external measures of performance. These comparative data might include industry best, best competitor, industry average, and appropriate benchmarks. Such data might be derived from independent surveys, studies, laboratory testing, or other sources.

Using the Five Types of Results

The types of measurable results examined by any company must represent the key effects on the company. Each of the five key types of indicators given above are related to each other and to the system within any business. When companies make changes to affect one or more of these indicators, the results affect the financial bottom line of the companies directly or indirectly. Thus it is critical that companies make the most knowledgeable decisions about what to initiate in order to achieve improved business results. A focus on business results encompasses the customer's evaluation of the company's products and services, the company's overall financial and market performance, and the results of all key processes and process improvement activities. Through this focus, the criteria's dual purposes—superior value of offerings as viewed by customers and the marketplace and superior company performance reflected in operational and financial indicators—are maintained. The business results should provide real-time information (measures of progress) for evaluation and improvement of processes, products, and services, aligned with overall business strategy.

The case studies in this book (Chapters 6–11) show how feedback from Baldrige-style assessments using the five types of results discussed above along with information from other sources can be integrated with theory (Baldrige and problem-selection matrixes; see Chapter 4) to create Knowledge (step 3 of the PDKA model). This Knowledge can then be prioritized and implemented through the business planning process to Act by deploying solutions and tracking results (step 4 of the PDKA model).

HOW TO GAIN RESULTS FROM KNOWLEDGE

Companies can gain results from business knowledge by converting all identified opportunities into knowledge, by prioritizing the knowledge before proceeding to develop action plans, and by acting to deploy the solutions to these opportunities into the business. With the PDKA approach, the company selects only the most important areas of opportunity for the next year. The areas of opportunity are linked to a problem-solving process and to a business-planning process, both of which should follow a Plan-Do-Check-Act logic. Again, the stumbling block for most businesses is their lack of a clear method and discipline in follow-through in their business planning, their problem solving, their workforce planning, and in most other parts of their business. Ironically, companies are willing to continue investing time in planning and gathering information relative to their businesses, usually in the form of an assessment or a survey of their businesses. However, when it comes to the critical follow-through of doing something with their feedback, the discipline and the know-how are missing. Sadly, companies are satisfied with stopping short of creating

Knowledge from their feedback data and setting Actions that could drive the improvement of their organization's entire business system, including their bottom line.

How can results actually be gained by Knowledge? The following key general actions will illustrate the basic sequence that needs to occur in order for a company to use its information to drive improvements and improve results.

1. Consolidate information sources.
2. Use Baldrige-style categories as an organizing framework.
3. Identify both strengths and weaknesses to improve.
4. Develop decision-making criteria by which to measure priority.
5. Attack the strongest strengths and the weakest weaknesses.
6. Determine the feasibility of implementing solutions.
7. Assign selected opportunity areas as tasks.
8. Select a sponsor and a team leader for each task.
9. Implement the improvements using a type of problem-solving process.
10. Integrate the actual improvements into the business planning process.

Consolidate Information Sources

Establish an action plan team (see Chapter 4) responsible for consolidating information sources into a feedback summary by integrating the Business Process Review feedback report with existing information sources, such as employee surveys, customer feedback, and past assessments as appropriate.

Use Baldrige-Style Categories as an Organizing Framework

Integrate all of this critical information by putting it into a *feedback summary matrix* (Chapter 4) that uses the seven Baldrige-style categories (Leadership, Strategic Planning, Customer and Market Focus, Information and Analysis, Human Resource Development and Management, Process Management, and Business Results) as the organizing framework to manage the information. At this point, information is being integrated with a sound theory (the seven Baldrige criteria) to achieve a new level of knowledge. Now, the action plan team is able to evaluate each piece of information (with some knowledge added) as either a strength area or an area of weakness.

Identify Both Strengths and Weaknesses to Improve

Determine the importance of each strength and weakness area by dividing these areas further into two more feedback summary matrixes, one for areas of weakness and one for strengths. The purpose of these matrixes is to determine those areas of opportunities (strengths and weaknesses are both considered opportunities) that, if improved, would help the business the most.

To become the best, businesses cannot only improve what they do poorly; they must also improve what they do well. To illustrate this theory, consider world-class table tennis. The sport is played throughout the world. Yet at every Olympics, it seems that players from the same two countries, Japan and China, are battling for the gold medal. The players from both nations have become world class, but their basic philosophies are different. The Japanese approach to being world class in table tennis is continuously to improve weaknesses; the Chinese approach is continuously to improve upon strengths. The logic holds that if a business were to improve not only its weaknesses, but also its strengths as well, it would increase its chances of becoming world class, as well as becoming financially sound. The key here is to choose only a few areas to improve, not hundreds. The PDKA model emphasizes the need to improve strengths as well as weaknesses to become truly competitive and to stay competitive.

Develop Decision-Making Criteria by which to Measure Priority

Develop sound business-related criteria to measure priority, which will determine which strength areas and opportunity areas are critical to address as quickly as possible to improve the company's bottom-line results. To make this determination, information must be prioritized, because all issues cannot be resolved at one time. First, the action plan team stratifies this data to eliminate duplication. Once stratification has occurred, the team must determine the priority of these strength and opportunity areas.

To assist in the prioritization of all opportunities, the team develops criteria by which to measure priority by using a tool called a *problem selection matrix* (see Chapter 4). The problem-selection matrix should have four to seven key criteria to address. Examples of such a criteria are impact on customer, impact on business bottom line, problem severity, a problem-solving team's ability to solve, to solve quickly, and to solve with low cost. For each opportunity area the criteria are scored on a scale from 1 (low correlation) to 5 (high correlation). The scores are totaled and priorities are now set by number. The highest point-getters are worked on. In other words, work at focusing your efforts on both the few areas of weakness that will improve the business the most and the one or two strengths that, if made better, would put it into a world-class category.

Attack the Strongest Strengths and the Weakest Weaknesses

The action plan team must now select the opportunities for implementation from areas of both weakness and strength. As we have noted, the criteria and problem-selection matrix yields a numeric scale that establishes a range of possible points that can be totaled for a potential improvement area. Areas of weakness whose scores fall below a certain percentage or areas of strength whose scores rise above a certain percentage should be considered critical. Thus, every deficiency and every strength that scores equal to or higher than the established percentage is attacked without question. Opportunity areas, either strengths or weaknesses, which fall below the established percentage may or may not be pursued at this time. At this point in the process, the information from feedback sources has been integrated with theory (in this case, the linking of the problem selection matrix with the Baldrige-style categories) to create knowledge. This level of knowledge can now be applied to predict, in Deming's words, "future outcome, with risk of being wrong, and . . . [that] fit[s] without failure observations of the past."[3] In other words, Knowledge can now be used to Act.

Determine the Feasibility of Implementing Solutions

Next, the action plan team uses the list of prioritized improvement opportunities to determine how many can be done feasibly within given time and dollar constraints. The team then schedules and budgets the selected opportunity areas and performs a simple cost/benefit analysis for management. The opportunity areas all have a projected scheduled completion and are tracked with actuals. Presentation and review dates are also noted on the schedule. The team should also estimate potential costs to implement a solution so that when it is time to implement, money has been budgeted. Once management reviews the proposed actions and the cost/benefit analyses, these tasks are factored into each organization's business plan actions, and tasks are assigned to deploy improvements.

Assign Selected Opportunity Areas as Tasks

The action plan team now assigns all selected opportunity areas as tasks, which can now be factored back into the business by using a problem-solving process or a more appropriate method to implement the opportunity. Some tasks can be improved another way: for example, if a task is already being worked on by a team, an organization is running a pilot on a problem, or a problem can be resolved by an individual or work unit. In these cases, the effort does not need to be duplicated or the problem solved with the problem-solving process. However, the existing efforts need to be tracked in the organization's business plans. For those opportunity areas that will be assigned as tasks, the action plan team now must select a sponsor, team leader, team members, and review team members.

Select a Sponsor and a Team Leader for Each Task

The sponsor should be a responsible manager who has ownership of the problem and who will take on the responsibility for helping the team succeed. The team leader is the driver of the task and will lead the team. The team leader will also rely on the sponsor to remove any roadblocks that come up and find ways to enable the team to continuously move forward. Task teams are improvement teams whose members are appointed by management, based on their expertise and willingness to be dedicated to the assigned problem or task through to implementation. The team will use problem-solving methods, as well as techniques such as brainstorming or root-cause analysis, to drive the problem to root cause.

Implement the Improvements Using a Type of Problem-Solving Process

Each task team uses the problem-solving process to attack the root cause of weaknesses or to pinpoint specific barriers to improving strength areas. Teams disband when the results are satisfactory and the process has been standardized. If the action plan team determines that task teams are not the appropriate way to implement this improvement, the team chooses a more appropriate method for implementing the improvement area, such as conducting a pilot, borrowing a world class process from another company, or having an individual or work unit come up with a solution.

Integrate the Actual Improvements into the Business Planning Process

The integrated list of strengths and weaknesses, the company's new knowledge, should be factored into its business planning cycle and tracked like any other critical activity. Status reports should be given at least monthly at business plan reviews. The status reports should update the status of the list of improvement initiatives, and indicators should be tracked to tell whether the implemented improvement is actually improving the bottom line. Management should continue to conduct business plan reviews that include the status of the task teams and the implementation of the solution for the opportunity areas. Once improvement opportunities have been implemented into the business and the problems have been solved and standardized throughout the rest of the company, the PDKA cycle has been successfully completed.

4

USING PDKA TO IMPLEMENT ASSESSMENT OPPORTUNITIES

INTRODUCTION

Chapters 2 and 3 stressed the discipline required to follow through with improvements. However, before companies can properly follow through, they must integrate the information they have gathered (surveys, focus groups, indicators, or assessments) with an organizing framework such as the Baldrige criteria. Prioritizing the knowledge gained from this integration of information and theory will enable them to implement the most critical opportunities first and, consequently, obtain the best results. Improvements such as increasing profits by 50% for five consecutive years, having an employee benefits package that has been repeatedly benchmarked as being best-in-class, losing zero customers in the last six years, and having customer satisfaction ratings that are the highest ever and higher than those of any competitors are the results associated with world-class companies. These are the results a company can obtain by using business knowledge from multiple information sources to drive business improvements. These are the results a company can obtain by using the PDKA process.

The PDKA process can best be explained at two levels, the *macro*, or overview of the process in terms of the key players required to make it work, and the *micro*, or more detailed, step-by-step methodology of the process.

THE PDKA MACRO PROCESS

The PDKA macro process has eight broad steps. Figure 4.1 shows the macro overview of the PDKA process.

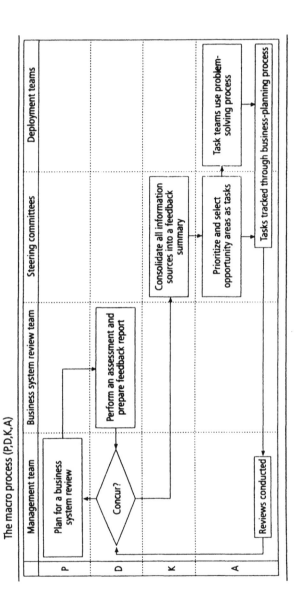

Figure 4.1 The macro process (PDKA).

The Teams That Drive the Macro Process

Five key groups of people who will drive information to knowledge and drive knowledge to bottom-line improvements are the Senior Management Team, the Business System Review Team, Steering Committees, and Deployment Teams, and Review Teams.

- *Senior Management Team* — This team, made up of the company's top executives, leads the business system review, concurs with how integration takes place, conducts reviews, and tracks the improvement tasks through the business planning process.
- *Business System Review Team* — This team performs a current assessment using an established criteria framework, such as the Malcolm Baldrige National Quality Award, and prepares the feedback report, which becomes the organizing framework for integrating all other information. It is composed of the company's business improvement specialists, appropriate executives, and the business planning team.
- *Steering Committees* — Steering Committees are those action plan teams that actually consolidate all information sources into the feedback report. They also prioritize and select opportunity areas (tasks) the company needs to attack to achieve bottom-line results. There is one steering committee for each category of knowledge, that is, the seven Baldrige categories (see Section 3.14 of this chapter). Each committee is composed of mid-level managers responsible for that area as well as key personnel.
- *Deployment Teams* — Deployment Teams are formed to find ways to improve each selected opportunity for improvement. These opportunities become tasks that are tracked. There is one deployment team for each task. Teams are composed of a sponsor from senior management, a team leader responsible for coordinating the task, and team members who will carry out assignments.
- *Review Teams* — Each Deployment Team is assigned a Review Team that will help it remove obstacles and get its solution implemented. A Review Team is a group of managers (perhaps someone from the senior management team), usually just two or three, who have a stake in the Deployment Team's success. The Review Team reviews the status of the Deployment Team's effort at agreed upon intervals, gives the team advice or suggestions and, most importantly, coaches and assists the deployment team and ensures that it succeeds.

The five types of teams work in concert to implement the PDKA process.

The Key Steps in the Macro Process

1. Leaders agree to review business systems.

The members of the Senior Management Team of an organization must understand and agree that they need to review their business systems in order to identify problem areas and to implement solutions that will drive improvement in their business.

2. Review Team assesses business systems.

The Senior Management Team in conjunction with the Business System Review Team assesses the current state of affairs of the business. Assessment, or the gathering of information by which to evaluate the condition of the business, can occur in many forms: employee surveys; supplier surveys; customer surveys; tracking of key indicators and the performance gaps in those indicators; focus groups for employees, customers, and suppliers; self-assessments; internal assessments or formal Baldrige-style assessments; and others.

While such information is routinely gathered under normal business operations, it is often gathered in disconnected ways and in an uncoordinated fashion. Often the feedback from the variety of information sources is never integrated into usable knowledge. The purpose of the Business System Review Team is to reassess the business in its current state with an assessment tool that may also be the framework by which to organize all other information and that incorporates all processes and systems within the business. A Baldrige-style assessment has proven to be a good assessment tool to use for this purpose. It is critical that the team members choose an assessment tool that will review the business in a holistic way and give them current information. This assessment tool will become the organizer by which to integrate all other prior and future information sources.

When the Business System Review Team has a new assessment done, it receives a feedback report from the assessors, and this report becomes the basis and the organizer for all other business information.

3. Leaders understand and support framework.

The Senior Management Team must fully understand the feedback report as well as all of the other information the company has. Then Senior Management Team members need to buy in to the comments and support this report as the basis by which to integrate all other business information. The objective is to use this report to focus the business on the biggest hitters so that the business will achieve bottom-line improvement.

4. Information is integrated into the framework.

This is the most critical step in the PDKA process, where a business integrates all of its information into the new assessment feedback report.

This is where the integration of information becomes new Knowledge for the business. Transforming information into business knowledge is critical and can only occur effectively if all random information is folded into an organizing framework, such as the criteria of the Baldrige Award. For example, no single assessment or survey will give an in-depth analysis about the business. In fact, because of human error, time constraints, or other issues, a single information source may give partial guidance toward some broad areas of improvement but not toward a definitive focus for significant business improvement. Only when a company brings in the specific detail from customer feedback sessions, detailed employee surveys, or other information sources does it really begin to obtain the true focus that is required to drive business improvement. The objective of the Steering Committees is to consolidate all of the information the business has on hand, integrate it into the organizing framework (i.e., the Baldrige criteria), then observe where the consolidation of this information leads them to focus their future efforts. They are turning speculation, subjective points of view, opinionated perspectives, and factual hard data into areas of common views from a variety of sources, which will lead them to a new organizational knowledge they never have had before. Sometimes this knowledge ends up being what they thought, but more often than not, this knowledge is vastly different from what any of them imagined.

5. **Prioritize and select opportunity areas.**

The Steering Committees must take all of this organizational knowledge, prioritize it based on a sound business logic, and select opportunity areas as tasks to deploy. The sound business logic is essentially a ranking criteria that helps an organization begin to stratify where it needs to focus its efforts in order to obtain the results it wants. The business logic needs to link to the underlying intentions of the strategic objectives of the organization. In other words, if a strategic objective is to improve profits by 20%, then part of the logic for deciding what to focus on needs to be based on whether it will help obtain this goal. Perhaps one of the strategic objectives is to grow a culture where employees want to stay and work their entire careers; if so, then the committees would ensure that the decision-making logic was designed so that the organization would focus on those critical areas that can obtain this objective.

6. **Implement tasks for opportunity areas.**

To implement the prioritized tasks, the organization will use a problem-solving process to determine the best solution to the identified area for improvement. Deployment Teams will go through the problem-solving process, identify the best solution, and then implement the solution.

7. **Track progress of tasks.**

The business, led by the Senior Management Team, tracks the progress of all improvement opportunity tasks. Tasks will occur at different stages and must be tracked, preferably through the business' planning process. This is assuming that the planning process has built-in review cycles every month or so. In other words, the Check step is occurring as it should.

8. **Review success of tasks.**

Reviews are conducted periodically on all tasks. These reviews are especially important for tasks that have been implemented so that the actual progress of the improvement is understood. If the implemented solution is working well, then the company needs to standardize this solution throughout; if the solution is not working as planned, then it needs to be revisited.

THE PDKA MICRO PROCESS: HOW TO DRIVE IMPROVEMENT

The following will explain a step-by-step methodology by which to implement the Knowledge and Act steps of the PDKA logic, as illustrated in Figure 4.2. Implementing these steps is the crux of how to use knowledge to drive improvement. Steps in the process preceded by an "A" in the following sections and in Figure 4.2 are actions; those preceded by a "D" are decisions.

Converting Information to Knowledge

Review Findings from the Feedback Report (Step A1)

It is very important that all information within a feedback report—whether it is from an assessment, an employee survey, a customer survey, or some other source of information—is clearly understood by the Senior Management Team. The information within the report typically addresses pluses and minuses within the organization. Often the feedback may imply something that is not popular, and it can become very easy to brush that piece of information off as erroneous feedback. It is critical for the Senior Management Team to review the feedback report—paying close attention to both the pluses and the minuses—and accept its use in improving the business. Too often organizations pay total attention to what is going poorly instead of focusing on both the weaknesses and the strengths. This report must be taken seriously, and it is important that the Senior Management Team not rationalize why certain comments are what they are.

The micro process (K,A)

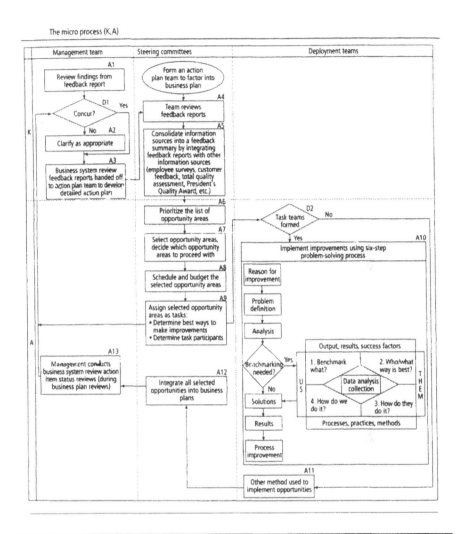

Figure 4.2   The micro process (KA).

They must accept them and integrate them into the other organizational information to form knowledgeable insight.

Obtain Senior Management Concurrence (Step D1)

The Senior Management Team must review the findings from the feedback report and make sure that it indeed understands and concurs with the findings. If the organization has used a sound process in gathering the information, then the Senior Management Team should accept the information as valid, even if it contradicts its own assumptions. Discipline is required here so that the information that is received is not watered down or tampered with, because that would not be in the best interest of the business' future success.

Modify the Feedback Report as Appropriate (Step A2)

For best results, feedback reports should not be modified at all, but if there is some key issue the entire Senior Management Team believes should be restated or changed because of political issues or any other issues they think are sensitive, then the report can be modified. Once managers clearly understand this feedback report and have no need to either clarify or make minor modifications due to the sensitivity of the information, then the report is considered final. The report is now ready to be used for focusing on the best improvement opportunities for the business.

Hand off the Report (Step A3)

Next, the feedback report is handed off to Steering Committees to develop detailed action plans around the feedback in this document. The reality of most businesses is that the Senior Management Team is composed of the *strategists*, not the doers. Therefore, to pretend that the Senior Management Team will actively push the action plans while they balance everything else they have going on will hinder success. The role of the Senior Management Team is to support this process, remove all roadblocks, and choose the best doers for implementing it. The *doers* are members of the Steering Committees. The Senior Management Team of the organization must make sure that the doers totally understand the content within the feedback report, and it must always be available to answer any questions or obtain any clarifications about the information in the feedback report. The Senior Management Team must also make sure that the doers are continuously recognized and acknowledged for their efforts to improve the business.

As noted earlier, Steering Committees are made up of selected managers and employees who will make certain that the PDKA process is followed through. There is one Steering Committee for each category of knowledge, that is, the seven Baldrige categories:

1. Leadership
   Alignment of fact with vision
   Social issue information
   Political issues
   Regulatory and legislative issues

2. Strategic Planning
   Benchmarking
   Competitive comparisons
3. Customer and Market Focus
   Customer information seminars
   Customer surveys
   Focus groups
   Databases containing detailed information on customers and their preferences
4. Information and Analysis
   Internet research and World Wide Web
   Internal self-assessment feedback
5. Human Resource Focus
   "Town meetings"
   Employee surveys and feedback
   One-on-one interviews
   Employee suggestion system
6. Process Management
   Alignment of measurement system with organization priorities
   Linking information with processes
   Control charts
   Process reviews
   Supplier feedback
   New product or service performance
   ISO 9000
7. Business Results
   Indicators that measure the business (employee safety, timely delivery, employee education, complaints, etc.)
   Results data showing how things are going
   Multi-year trend data
   Customer Satisfaction Index

Review the Feedback Report (Step A4)

Members of the Steering Committees (both managers and employees) review the feedback report to make sure that they understand the content. Again, the Senior Management Team will help the Steering Committees understand what is in the feedback report and will be readily available to clarify or answer further questions. However, at this point the baton is essentially handed off to the doers, and these people are being relied on by the Senior Management Team to help them achieve bottom-line improvements to the business.

Consolidate information sources							
							How To:
Feedback summary - Opportunity areas							
	Information sources						Need for Improvement (points)
Comment Area	A	B	C	D	E	Remarks	
Senior managers have researched (benchmarked), developed, and are actively leading the deployment of improvements throughout the organization	X	X	X	X	X		5
Managers throughout the organization participate in communicating and reinforcing quality values	X			X	X		4
Performance _(...)_ and supervisors _(...)_ business planning _(...)_ quarterly							1
Strong evide _(...)_ the entire or _(...)_ across						Need for improvement 5Xs = 5 pts 4Xs = 4 pts 3Xs = 3 pts 2Xs = 2 pts 1Xs = 1 pt	3
There is a clear focus on _(...)_ ing all corporate activities to meet our strategic goals							3
Top management provides significant resources for learning and recognition to support and reinforce employee contributions			X				1
A committed core of employees with learning experience have bought into and support management's quality approach	X		X		X		3

List all Comments here

Put a Check in the box if a comment was stated in that information source.

Figure 4.3   Consolidate information sources (how to).

Consolidate Feedback Resources and Integrate Them with Other Sources (Step A5)

Members of the Steering Committees are responsible for consolidating information sources into a *feedback summary matrix* by integrating the feedback report with existing information sources in their organization, such as employee surveys, customer feedback, past assessments, and others as appropriate. At this point in the explanation of the PDKA process, it is assumed that the core feedback report used is a Baldrige-style assessment feedback report divided into the associated seven categories noted above. Even if the company has not had a Baldrige-style assessment performed, it could still use the seven categories as the organizer for all of the information the business currently has. In the case of non-U.S—based companies that use the European Quality Award assessment process, the Deming Prize process, the Peacock Award process, or other process, the basic organizing criteria for those awards can be used as the organizing framework by which to integrate all of the company's information.

Figure 4.3 shows how to lay out a feedback summary matrix to consolidate different types of information. In the figure, the comment area of the matrix is where all of the feedback from various sources is listed. The A, B, C, D, and E columns could represent five different feedback sources (for example, employee surveys, customer surveys, self-assessments, state award assessments, and national award assessments). If the comment was identified in the feedback of that information source, an X is placed in the box under Information Sources. The X's in the columns are added to give a quantifiable score for each comment. This step is the beginning of the integration and prioritization process.

Steering Committee members continue the integration and prioritization process by determining the importance of each area of strength and weakness by further dividing the areas into two matrixes, one for areas of weaknesses and one for areas of strengths. For example, there may be a matrix that lists all of the strengths in Baldrige Category 3, Customer and Market Focus, and there may be a matrix that lists all of the weaknesses in Category 3. The purpose of these matrixes is to determine those areas of opportunities which, if improved, would help the business the most. As we have already established, to become the best, businesses cannot only improve what they do poorly; they must also improve what they do well. Think back to the Japanese and Chinese table tennis team philosophies described in Chapter 3: that is, the Japanese focus on improving weaknesses, and the Chinese focus on improving upon strengths. The key here is to choose only *a few areas* to improve, not hundreds. To do this, an organization must prioritize, develop action plans, implement the plans, and review.

Figure 4.4 shows an example of feedback on weaknesses in the Leadership, Baldrige Category 1. An X appears under each information source that acknowledges that comment as a weakness. Adding the X's gives the total points under the Need for Improvement column.

Figure 4.5 follows the same logic for strength areas in the Leadership category.

At this point, members of the Steering Committees have integrated all of the organization's information into a single collection of information that has been organized by the seven key business criteria associated with the Baldrige criteria. The information has essentially been broadly categorized into natural groupings (i.e., Leadership, Strategic Planning, Customer and Market Focus, Information and Analysis, Human Resource Development and Management, Process Management, and Business Results) based on subject matter content. The Steering Committees have begun to integrate and prioritize the masses of information, but they have not yet begun to focus on those that will attain the best results.

To accomplish this task, they must develop criteria to measure priority among all of the pieces of information. A mistake that many companies

Consolidate information sources							
							Example:
Feedback summary - Weakness areas:							
Comment Area	Information sources					Remarks	Need for Improvement (points)
	Employee Survey	Customer Survey	Self- Assessment	State Award Assessment	National Award Assessment		
Limit process measures used	X				X		2
Few trends and current levels of performance are shared	X	X		X	X		4
Industry averages used not "best in class" (6.1)		X					1
No benchmark comparison provided related to supplier quality results		X				Need for improvement 5Xs = 5 pts 4Xs = 4 pts 3Xs = 3 pts 2Xs = 2 pts 1Xs = 1 pt	1

Figure 4.4   Consolidate information sources (weaknesses) (example).

make is that they try to tackle everything, usually at the same time. These efforts result in frustration, overworked employees, and usually minimal, if any, improvement to the business, so the criteria will help prioritize the lists of improvement opportunities. In other words, these criteria will determine which strength areas and which areas of weakness should be improved first.

Prioritize the Opportunity Areas (Step A6)

At this point in the process, the organizations now have a consolidated list of all of their feedback. This step is necessary but not yet complete. The feedback areas must now be further prioritized, because all comments cannot be resolved at one time. Criteria that will assist in the prioritization must be developed. Essentially, these criteria will determine which strength areas should be improved first and which areas of weakness should be improved first. To do this, the Steering Committees first stratify the feedback to eliminate duplication. In other words, sometimes comments are similar if not identical. It is important that the committees consolidate the similar information so that it does not add confusion or result in duplicate actions later.

Once stratification has occurred, the Steering Committees must determine the priority of these strength and weakness areas. To assist in the prioriti-

Consolidate information sources

Example:

Feedback summary - Strength areas

Comment Area	Information sources					Remarks	Need for Improvement (points)
	Employee Survey	Customer Survey	Self-Assessment	State Award Assessment	National Award Assessment		
Senior managers have researched (benchmarked), developed, and are actively leading the deployment of improvements throughout the organization	X	X	X	X	X		5
Managers throughout the organization participate in communicating and reinforcing quality values	X	X		X	X		4
Performance requirements for managers and supervisors are established through the business planning process, reviews are conducted quarterly	X						1
Strong evidence of being a good neighbor across the entire organization	X	X			X	Need for improvement 5Xs = 5 pts 4Xs = 4 pts 3Xs = 3 pts 2Xs = 2 pts 1Xs = 1 pt	3
There is a clear focus on aligning all corporate activities to meet our strategic goals		X	X	X			3
Top management provides significant resources for learning and recognition to support and reinforce employee contributions			X				1
A committed core of employees with learning experience have bought into and support management's quality approach	X		X		X		3

Figure 4.5   Consolidate information sources (strengths) (example).

zation of all opportunities, the committees can develop criteria by which to measure priority by using a tool called a *problem selection matrix* (see Figures 4.6–4.8). It is important that the committees give some significant thought to what the criteria should be. The criteria should be based on what is critical and important to the organization. The criteria should also be based on those things that will lead the organization to achieving results that improve the bottom line. It is also important that the criteria be made up of only a handful of dimensions; more than six criteria dimensions are probably too many. When the team decides on the best criteria to use as the prioritization process, its next step is to put those criteria into a problem-selection matrix so that prioritization can begin. The problem selection matrix that is shown in Figures 4.6, 4.7, and 4.8 has the following six criteria:

1. Impact on the customer.
2. Impact on the business bottom line.
3. Problem severity (need for improvement).
4. Team's ability to solve.
5. Team's ability to solve quickly.
6. Team's ability to solve with low cost.

	Prioritize opportunity areas						
Problem selection matrix (scale 1, 2, 3, 4, 5 (1 = low, 5 = high)						How To	
Comment Area	Impact on Customer	Impact on Business Bottom Line	Problem Severity	Team's Ability to Solve	Team's Ability to Solve Quickly	Team's Ability to Solve with Low Cost	Total Score
Senior managers have researched (benchmarked), developed, and are actively leading the deployment of improvements throughout the organization	1	5	5	5	4	4	24
Managers throughout the organization participate in communicating and reinforcing quality values	3	3			3	3	21
Performance requirements for managers and supervisors ...d through the busin... ...reviews are con...	1	5				4	Add scores to obtain totals
Stro... neighb... organ...	1		Put in a Score of 1-5 in each box			5	
There is a ... corporate activitie... ...meet our strategic goals	2	3			5	5	23
Top management provides significant resources for learning and recognition to support and reinforce employee contributions	2	4	1	2	2	3	14
A committed core of employees with teaming experience have bought into and support management's quality approach	1	4	3	2	2	3	15

(annotations: "List all Comments here")

Figure 4.6   Prioritize opportunity areas (strengths and weaknesses) (how to).

For each opportunity area (strengths and weaknesses), these six criteria are scored on a scale from 1 to 5. A score of 1 is low correlation; 2 is somewhat low correlation; 3 is moderate correlation; 4 is somewhat high correlation, and 5 is high correlation. So, each piece of information in the matrix will have a score of 1 to 5 associated with each of the six criteria. Figure 4.6 shows how to prioritize opportunities by first listing all comments, scoring each comment as it affects each criteria, then totaling the score.

For example, the first criteria (Impact on Customer) may score a 5 for a particular opportunity listed in the Comment Area of the matrix. This means that if the opportunity were implemented, it would have a significant positive impact on the customer, or it could mean that if the opportunity were left alone, it could have a significant negative impact on the customer. The second criteria (Impact on Business Bottom Line), may score a 3 for that opportunity. This means that if the opportunity were implemented, it would have just moderate effect on the bottom line of the business. The criteria of Problem Severity scores a 4. This criteria is simply the numbers that are added in the column "Need for Improvement" in the feedback summary matrix shown in Figures 4.3, 4.4, and 4.5. The logic for including this information is that if the same comments

Prioritize weakness areas							
Problem selection matrix (scale 1, 2, 3, 4, 5 (1 = low, 5 = high)						Example	
Comment Area	Impact on Customer	Impact on Business Bottom Line	Problem Severity	Team's Ability to Solve	Team's Ability to Solve Quickly	Team's Ability to Solve with Low Cost	Total Score
Limit process measures used	1	5	5	5	4	4	24
Few trends and current levels of performance are shared	3	3	4	5	3	3	21
Industry averages used not "best in class" (6.1)	2	3	1	5	5	5	21
No benchmark comparison provided related to supplier quality results	2	4	1	2	2	3	14

Figure 4.7   Prioritize weakness areas (example).

have occurred in multiple sources of information, the problem is certainly highly observed and may be severe. With a score of 4, this criteria is a significant problem for the organization, one that probably permeates throughout the whole organization. The criteria of Deployment Team's Ability to Solve scores a 5, which means that a task team of employees could analyze this opportunity and find a way to make it significantly better. The criteria of Deployment Team's Ability to Solve Quickly scores a 1, which means that although the team will be able to come up with a solution, it will take a long time to implement. The criteria of Deployment Team's Ability to Solve with Low Cost scores a 3. This means that it can be implemented within a reasonable cost; it will not be outrageously expensive, but it will not be cheap. So, for this particular opportunity area in the example, the total score is 5 + 3 + 4 + 5 + 1 + 3 = 21 points. All other opportunity areas should be scored with the same correlation. The scores for all of the opportunities are totaled, and priorities are now determined by a quantified number. So, based on logic, the opportunities with the highest point totals should be the ones that the organization needs to work on first. Both the areas of strength and the areas of weakness should be worked on for the highest point-getters. Figures 4.7 and 4.8 show examples of how the listed comments are quantified and thus prioritized.

			Prioritize strength areas				
Problem selection matrix (scale 1, 2, 3, 4, 5 (1 = low, 5 = high)						Example	
Comment Area	Impact on Customer	Impact on Business Bottom Line	Problem Severity	Team's Ability to Solve	Team's Ability to Solve Quickly	Team's Ability to Solve with Low Cost	Total Score
Senior managers have researched (benchmarked), developed, and are actively leading the deployment of improvements throughout the organization	1	5	5	5	4	4	24
Managers throughout the organization participate in communicating and reinforcing quality values	3	3	4	5	3	3	21
Performance requirements for managers and supervisors are established through the business -planning process; reviews are conducted quarterly	1	5	1	5	4	4	20
Strong evidence of being a good neighbor across the entire organization	1	3	3	5	5	5	22
There is a clear focus on aligning all corporate activities to meet our strategic goals	2	3	3	5	5	5	23
Top management provides significant resources for learning and recognition to support and reinforce employee contributions	2	4	1	2	2	3	14
A committed core of employees with teaming experience have bought into and support management's quality approach	1	4	3	2	2	3	15

Figure 4.8   Prioritize strength areas (example).

The problem-selection matrix for setting priorities is linked to the Baldrige scale of the Malcolm Baldrige National Quality award, which rates a company's maturity in a given area on a scale of 0% (immature) to 100% (sound, systematic, and fully-fact based approach). The cutoff points for the problem selection matrix are set at 30% and 70% on the Baldrige scale. What this means is that areas of opportunity whose scores fall below 30% (21 points) or areas of strength whose scores rise above 70% (21 points) on the Baldrige scale are considered critical. Thus, every area of weakness and every area of strength that scores equal to or higher than 21 points is attacked without question. Any opportunity area, whether a strength or weakness, that falls below a score of 21 points may or may not be pursued at this point. The organization now has a series of matrixes, two sets of problem selection matrixes for each of the seven Baldrige categories, one set showing the strength areas in priority order for each category (Figure 4.8), and one set showing the weakness areas in priority order for each category (Figure 4.7). In total, the organization should have 14 separate problem selection matrixes, all of which have opportunity areas prioritized from a score of 0 to 30. The organization proceeds with the highest-scoring opportunity areas regardless of which matrix they are on.

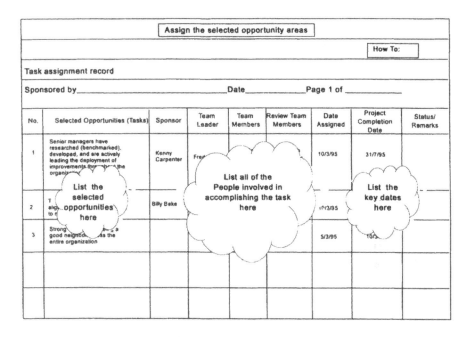

Figure 4.9   Assign the selected opportunity areas (how to).

At this point in the process, the information from feedback sources has been integrated with theory (in this case the linking of the problem-selection matrix with the Baldrige categories) to create knowledge. This level of knowledge can now be applied to predict, in Deming's words, "future outcome, with risk of being wrong, and . . . [that] fit[s] without failure observations of the past."[3] In other words, the Knowledge that is represented in the form of prioritized opportunities can now be used to *Act*, or to implement in order to actually improve the results of the business.

Acting upon Knowledge to Improve the Bottom Line

Select and Proceed with Opportunity Areas (Step A7)

The Steering Committees must now select the areas of weakness and strength to be implemented. Again, the committees have followed a disciplined logic that allows all the opportunity areas to be scored consistently. The scores for all of the opportunities lead to prioritization among the opportunity areas. So, based on logic, the opportunities with the highest point totals will be the ones that the organization works on first.

Assign the selected weakness areas							
						Example	
Task assignment record							
Sponsored by_____Date_____Page 1 of _____							
No.	Selected Opportunities (Tasks)	Sponsor	Team Leader	Team Members	Review Team Members	Date Assigned	Project Completion Date	Status/ Remarks
1	Few trends and current levels of performance are shared	Billy Baker	Brian Butcher	Green Lantern Super Girl The Avengers		10/3/95	30/5/95	
2	Limited process measures used	Peter Plumber	Emma Royd	Plant Reps		1/5/95	28/10/95	

Figure 4.10   Assign selected weakness areas (example).

The Steering Committees select the top three to five point-getters from among all of the prioritized feedback (new knowledge) and establish key dates and key people for the tasks to be assigned. Figure 4.9, a Task Assignment Record, shows the selected opportunities, along with the people who will improve and implement the opportunities. Figures 4.10 and 4.11 show examples of the assigned weakness and strength opportunities.

Schedule and Budget the Selected Opportunity Areas (Step A8)

In order to actually improve any opportunity area, discipline must be continued in the form of maintaining a schedule and a budget, just like any other project that an organization may implement. Figure 4.12, an Opportunity Area Planning Worksheet, shows a schedule of selected opportunity areas.

Money must be budgeted into the business planning process for the Deployment Team's effort to come up with a solution as well as for the actual cost to really deploy the solution into the organization. So, the Steering Committees use the list of prioritized improvement opportunities to determine how many can be feasibly done within given time and dollar constraints. The committees schedule and budget the selected opportunity areas and perform a basic cost/benefit analysis for the Senior Management Team. The simple cost/benefit analysis is the last validation that a particular

	Assign the selected strength areas							
							Example	
Task assignment record								
Sponsored by_____Date_____Page 1 of _____								
No.	Selected Opportunities (Tasks)	Sponsor	Team Leader	Team Members	Review Team Members	Date Assigned	Project Completion Date	Status/ Remarks
1	Senior managers have researched (benchmarked), developed, and are actively leading the deployment of improvements throughout the organization	Kenny Carpenter	Fred Farmer	Ron Morris Fred Smith Bill Brown	Jacky Smith Henry Martin Bob Roberts	10/3/95	31/7/95	
2	There is a clear focus on aligning all corporate activities to meet our strategic goals	Billy Baker	Brian Butcher	Green Lantern Super Girl The Avengers		10/3/95	30/5/95	
3	Strong evidence of being a good neighbor across the entire organization					05/03/95	10/3/95	
4	Managers throughout the organization participate in communicating and reinforcing quality values	Christopher Cook	Emma Royd	Plant Reps		03/10/95	10/3/95	

Figure 4.11   Assign selected strength areas (example).

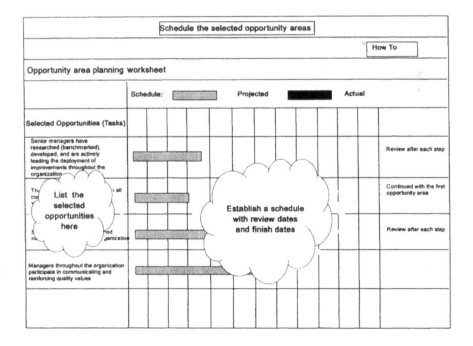

Figure 4.12   Schedule the selected opportunity areas (how to).

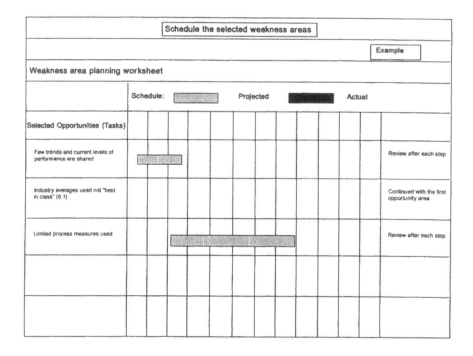

Figure 4.13   Schedule the selected weakness areas (example).

opportunity is indeed worth doing. The Steering Committees also estimate potential costs to implement a solution so that when it is time to implement, money has been budgeted. The opportunity areas all have a projected scheduled completion date and are tracked with actuals. The status of the implementation should be checked periodically, preferably at least once a month. Presentation and review dates are also noted on the schedule. Once the Senior Management Team reviews the proposed actions and the cost/benefit analyses, these tasks are factored into each organization's business plan actions, and tasks are assigned to deploy improvements. Figures 4.13 and 4.14 show schedules of selected opportunity areas, i.e., selected weakness areas and selected strength areas.

Assign Selected Opportunity Areas as Tasks (Step A9)

The Steering Committees now assign all selected opportunity areas as tasks that can be factored back into the business by using a problem-solving process or some other established method that may be more appropriate for an organization to implement the opportunity. Some tasks

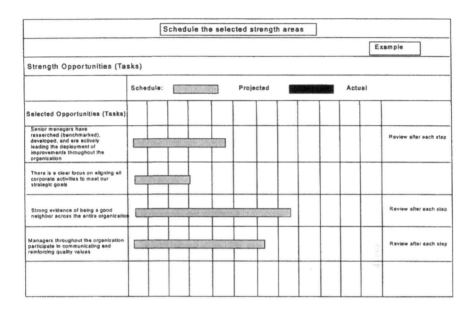

Figure 4.14   Schedule the selected strength areas (example).

may not require a standard problem-solving process; some tasks can be improved other ways. For example, a team may already be working on a task; an organization may be running a pilot on a problem; or a problem may be resolved by an individual or work unit. In these cases, the effort does not need to be duplicated or the problem solved with the formal problem-solving process. However, the existing efforts also need to be tracked and linked to the organization's business planning process.

Determine Whether to Form Task Teams (Step D2)

The Steering Committees will decide whether it is necessary to form Deployment Teams. For those opportunity areas that will be assigned as tasks, appropriate team members and a team leader will be selected. The team leader has responsibility for coordinating the task; the team members work together to accomplish the task. It is important that the team members have the right skill sets to solve the particular problems; likewise, each team member must have the dedication and interest in the opportunity to actually see it through to implementation. These will be the individuals who take on the responsibility of coming up with the way to improve the opportunity, as well as the best way to actually implement the solution throughout the organization. The Steering Committees also select a sponsor and review team members. The sponsor is a senior manager who has ownership of the opportunity area and serves to assist the team in implementing recommendations. The Review Team is a group of managers who coach the Deployment Teams and give them suggestions periodically through reviews.

Implement Improvements Using a Problem-Solving Process (Step A10)

Each Deployment Team uses a problem-solving process to attack the root cause for each area of opportunity. A problem-solving process offers a consistent and disciplined method each task team can use, which eventually reduces cycle time and enables teams to complete tasks faster and more effectively. When an opportunity has been solved and the solution has been implemented, the implemented solution is checked. If it actually does what the team thought it would do, the team disbands. In other words, the team stays together until the results are satisfactory and the process has been standardized.

Use Another Method to Implement Opportunities (Step A11)

If the Steering Committees determine that Deployment Teams are not the appropriate way to implement a particular improvement opportunity, they make sure that an alternative method is followed. The Steering Committees determine the more appropriate method for implementing the improvement area, such as conducting a pilot program, borrowing a world-class process from another company, or having an individual or work unit come up with a solution. Regardless of the method used, when the opportunity has been solved, implemented, checked, and approved, the task is finished and the process is standardized.

Integrate Selected Strengths and Opportunities into Business Plans (Step A12)

Prioritized opportunities must be treated as critical to the success of the business and critical to improving the bottom line of business. In most businesses, critical business issues are tracked through the organization's business planning process. The knowledge that the organization now has is indeed critical and vital to the future success of the company and should be tracked in the business plan. *All* opportunities, both the strength and weakness areas, must be integrated into the business planning process. Once again, it is very tempting to focus only on the weaknesses, but the strengths must be improved as well, so that the business can become world-class and highly competitive. Once strength and weakness opportunity areas have been integrated into business plans and their progress tracked, the PDKA cycle has been successfully completed, and the company will begin seeing indications of better results.

Conduct Reviews (Step A13)

Finally, the Senior Management Team conducts business reviews, which include the status of the tasks being worked on by Deployment Teams and the implementation of the solutions for the opportunity areas. These business reviews should be linked with the organization's business plans, and the updating of the business plans.

CONCLUSION

The process of improving a business and driving the company toward world class can be simple if discipline, clear logic, and action are followed. Again, the issue that most businesses face is that they do not create knowledge; they do not prioritize; they do not determine how to implement; and they do not follow through. Using PDKA logic to review a business system can provide an organization with invaluable knowledge related to the comprehensive performance level of its operations. This process allows an organization a chance to grow and learn about its business operations, determine how its efforts can be focused on systematic improvement, and experience the satisfaction of actually seeing improved business results.

5

HOW WORLD-CLASS COMPANIES DEMONSTRATE THE PRINCIPLES OF THE PDKA PROCESS

INTRODUCTION

The PDKA process is all about using knowledge in a disciplined way for business improvement. As an examiner of quality award applicants, as a manager who led his organization through the process of becoming world-class, as a student of the quality process, as a consultant who has helped other companies win quality awards, and as a contributor to the quality award process itself, the author of this book derived the PDKA process through years of thought, testing of hypotheses, and positive outcomes. Evolved from the PDCA model, the PDKA model goes beyond information and integrates information with theory to create knowledge.

As we have seen, the PDKA process is an entire system providing a logic and framework with which to assess and build upon both strengths and weaknesses. It is a process that integrates process improvements into the very fabric of the organization so that selected process improvements are scheduled and funded through an organization's business planning and budget process. The process requires buy-in from senior management and implementation by the employees who perform the tasks. Companies who have used the entire PDKA process have seen dramatic improvements. Companies who have used key ingredients, which are part of the PDKA logic, have seen improvements as well.

While the PDKA process is a well-thought-out system that addresses every facet of the organization, it is not widely known. However, in reading the case studies, one can see that the principles of the PDKA process permeate the companies' movement toward world-class status. Of the companies whose case studies have been included in this book, two have implemented the PDKA process as designed and have found the results to be significant (the United Way of Middle Tennessee and Fossil and Hydro Power of the Tennessee Valley Authority). The United Way of America Excellence in Service Quality Awards has modeled a learning tool based on the PDKA process. The tool assists service sector not-for-profits in improving their organizations.

The fact that most or all of these companies have participated in a quality awards application and assessment of some type indicates that they have an organizing framework or a theory by which to integrate their information to turn it into knowledge. How well they have done this may vary in that they may not have used an approach as disciplined and orderly as that of the PDKA process. For example, they may not use the matrixes to cross-index their feedback assessment with a problem-selection matrix to help prioritize those strengths or weaknesses that must be attacked first, or they may not have a rigid decision-making criteria, or they may not have a formal problem-solving process by which to implement opportunities.

Success of the PDKA process lies in its logic and mathematical precision. Information is categorized by quality award criteria. It is analyzed, itemized, and prioritized. Matrixes are used to mathematically and objectively determine those items that must be implemented, whether they be strengths or weaknesses. The process follows through with how the items will be implemented (problem-solving and schedule) and how they will be paid for (budget). Finally, it ensures that the items have been implemented successfully and that their success is replicated throughout the business. The process is logical and thorough and successful.

THE VALUE OF CASE STUDIES

How the Case Studies Were Selected

The case studies that have been selected show world-class methodologies that parallel 1 or all of the 12 principles of PDKA (i.e., executive buy-in, action plan teams, consolidation of information, an organizing framework, identification of strengths and weakness, clear decision-making criteria, prioritization of strengths and weaknesses, assignment of tasks and leaders, problem-solving process, progress tracking, periodic reviews, and standardization; see Chapter 2, Section 3.4). Some of the case studies focus on one or a few processes that happen to be information-related (such as customer surveys or employee feedback) and actually follow much of the PDKA logic. Some of the case study organizations discuss their entire business and the key methods that allow their business to achieve significant improvements. Some of the case study organizations explain their history of quality improvements; some lay out their process flowcharts; and others explain lessons learned about what has worked well and what has not. And some of the case study organizations have consulted with the author and actually use the PDKA logic as described in the previous chapters. All of the case study organizations provide insight into the 12 principles of PDKA and provide explanations of how they drive improvements.

In summary, the case studies in this book have been chosen because the companies

- Participate in total quality awards programs that use a performance criteria.
- Are world-class companies that have achieved significant positive results and are dedicated to improvement.
- Actively improve their business systems to drive results.
- Are diverse; represent government, private, and not-for-profit sectors; vary in size and geographic location (U.S. and international); offer a variety of products and services.
- Are practicing many of the 12 principles of PDKA (i.e., using multiple sources of information and trying to integrate together to give further insights). However, the case study organizations are at various points of implementing the use of the PDKA principles.
- Are well known or are destined to be well known.
- Have demonstrated significant improvement at least partially due to implementing opportunities based on knowledge.

How the Case Studies Reinforce the PDKA Process

By reviewing the case studies, readers can observe variations on the themes of PDKA and can recognize the approaches taken by different industries and the processes as they apply to both service- and product-based companies.

Readers can also see

- How simple the PDKA process can be. PDKA is a disciplined and process-oriented approach, but it is simple. Complexity in a process may cause organizations not to implement effectively and thus may lead the process to fail.
- How PDKA is not a separate business activity but one that is meshed with planning and builds on existing information. If PDKA is made

to stand alone as a separate activity, then it risks being seen as a side activity that is really not very important.

■ How to implement the PDKA process.

■ How to learn from examples of the process flowcharts using PDKA logic and from the tables and discussions of types of information used to create strategic knowledge. The case studies clarify how strategic knowledge occurs by showing and explaining the tools and techniques that bring the process to life. Many of the case studies describe step by step how organizations implement their methodologies.

■ How there may be similarities between a certain case study and the reader's company and how the reader may observe some approaches that could work in his or her company. While not all of the case studies implement the whole PDKA process, all of the organizations that contributed case studies make use of the principles behind the PDKA logic. These principles and methods explained in the case studies parallel what many organizations are currently doing or wanting to do. Perhaps some of the key principles and methods explained in the case studies may be seen as similar and thus picked up and used. However, because the entire PDKA process can be easily adapted in any company, it is most beneficial for companies to adapt the process in its entirety.

■ How to implement a method that could be used by their company. It is important not to just observe a process and think of it as a good idea; it is also important to be able to understand the nuances of the process and to learn how to implement it. Learning occurs best when something is explained precisely and examples of how others implemented the process are used. But the best learning will occur by an organization that implements the process for itself.

■ How they may gain further detailed insights about PDKA based on the case study descriptions. There are multiple steps to the PDKA process flowchart, and there are 12 principles to the PDKA logic. The importance of each of the steps is explained in detail as well as how the step can be accomplished. Detailed insights occur when one understands the depth of each step and its importance in the process.

LEARNING FROM THE CASE STUDIES

While the case studies reveal the major successes of the companies that have strived for world-class status, they also teach some valuable lessons.

Choosing the Right Assessment Tool

Although the major example of an assessment tool used in the first four chapters of this book is the Malcolm Baldrige National Quality Award, it is by no means the only such tool. As the case studies show, many quality awards contain the criteria to review the business in a holistic way. The most important aspect of the selection of an assessment tool is that it will become the organizer by which to integrate all other prior and future information sources; therefore it should closely parallel the strategic direction of the organization.

Using Multiple Information Sources

No single assessment or survey will give an in-depth analysis about the business. In fact, because of human error, time constraints, or other issues, a single information source may give partial guidance toward some broad areas of improvement but not toward a definitive focus for significant business improvement. While several of the case study organizations in this book concentrated on only one or two major sources, others show that the most effective method of using organizational information is to integrate all information to form a more comprehensive strategic knowledge base.

Avoiding Rationalizing Unpopular or Negative Findings

Senior managers of companies who are new to the assessment process often feel that the feedback implies something negative, and they may want to brush that piece of information off as erroneous feedback. It is important that the Senior Management Team not rationalize why certain comments are what they are. The case studies show that those in senior management should not impose their own personal opinions concerning whether the information is accurate or not, but should accept it as part of the assessment process and learn from it.

Trying Not to Correct All Problems at Once

Those organizations that choose only a few areas to improve at one time, that prioritize, develop action plans, implement the plans, and review, make the most significant progress. The studies reinforce the lesson that doing too many things at once oftentimes leads to chaos and results in accomplishing next to nothing.

Focusing on Strengths as Well as Weaknesses

Weaknesses are typically highlighted, and it is important to identify and improve weaknesses. However, many organizations ignore their strengths. If they continue to ignore their strengths, then their strengths will either level out or they may backslide to mediocrity. For this reason, it is important for an organization to push its strengths to be best-in-class. The case studies represent some best-in-class strengths. Oftentimes because of the systematic relationships within organizations, if an organization improves its strengths, those improvements may actually cause some of the weaknesses to improve as well.

Involving Employees

The case studies show that organizations that involve their employees are more successful. Employees should be involved in rounding up organizational information and should be involved in aggregating the information into knowledge. Employee involvement increases employee trust within any organization, and employees have a tremendous amount of organizational intelligence they need to share. Employees should also be involved in implementing improvements into the organizations.

Choosing The Right Criteria For Prioritizing

The case studies demonstrate that organizations should give some significant thought to what criteria should be used to prioritize strengths and weaknesses. The criteria should be based on what is critical and important to the organization and on those things that will lead the organization to achieving results that improve the bottom line. Also, the criteria should be made up of only a handful of dimensions. The prioritized improvement opportunities drive the determination of how many feasibly can be done within given time and dollar constraints. Without clear decision-making criteria, companies perhaps risk working on the wrong things at the wrong times, or even working on conflicting initiatives.

Planning and Budgeting for the Strengths/Weaknesses To Be Improved

Improvement processes cannot be implemented effectively if they are not integrated into the business planning process. Cost–benefit studies along with problem-solving processes and teams take time and money. Therefore, money must be budgeted into the business planning process for the Deployment Team's effort to come up with a solution as well as for the actual cost to deploy the solution into the organization.

Completing the Process

Successful organizations must implement improvements, review those improvements, and continue improvement efforts. Additionally, organizations must not only take the steps to implement improvements but also standardize successful implementations for greater improvement. Successful companies demonstrate that the process is not complete until the results are satisfactory and the process has been standardized.

THE CASE STUDY COMPANIES

There are 15 case studies represented by 15 different organizations that make up a cross section of industries from different parts of the world. The case studies have been grouped together in Chapters 6–11 as follows:

- Chapter 6, The Quality Award Programs
  The President of the United States Quality Award Program (PQAP)
  The United Way's Excellence in Service Quality Award Program (ESQA)
  The Northern Ireland Quality Award Program (NIQA)
- Chapter 7, The Private Manufacturing Sector
  Milliken Denmark Ltd.
  Pal's Sudden Service
  Ulster Carpet Mills Ltd.
- Chapter 8, The Private Service Sector
  BI Performance Services
  British Telecom Northern Ireland (BTNI)
- Chapter 9, The Public and Not-for-Profit Service Sector
  Royal Mail
  The United Way of Middle Tennessee (UWMT)
- Chapter 10, The Government Service Sector
  Federal Supply Service Northeast and Caribbean Region (Fed. SS)
  Tennessee Valley Authority: Fossil & Hydro Power (TVA F&HP)
  National Aeronautics and Space Administration: Kennedy Space Center (NASA)
- Chapter 11, The Military Service Sector
  Red River Army Depot
  U.S. Army Armament Research, Development & Engineering Center: Picatinny Arsenal (USARDEC)

Table 5.1 Case Study Variety

	Manu-facturing	Service	Private	Govern-ment (Public)	U.S.-Based	Interna-tionally Based	Not-for-Profit
BTNI		×	×			×	
BI Perf. S.		×	×		×		
Fed. SS		×		×	×		
Milliken	×		×			×	
NASA		×		×	×		
NIQA		×	×			×	
Pal's S.S.	×	×	×		×		
PQAP		×		×	×		
Red River		×		×	×		
Royal Mail		×		×		×	
TVA F&HP		×		×	×		
Ulster	×		×			×	
ESQA		×			×		×
UWMT		×			×		×
USARDEC		×		×	×		

Case Study Variety

Table 5.1 lists each organization and depicts the variety within the companies, which represents as well the variety you can expect to find in reading their specific case studies.

PDKA Matrix

Although many companies and organizations represented by the case studies have incorporated many or all of the 12 principles of PDKA, Table 5.2 matches the case studies to the salient principles.

Of the companies and organizations represented by the 15 case studies, most emphasized consolidation of information, an organizing framework, decision-making criteria, and periodic reviews. In fact, the companies and organizations represented by the case studies may have implemented many other facets of PDKA but may not have included them in the case studies or may have not emphasized them.

The principles least emphasized in the case studies are action plan teams, prioritization of strengths and weaknesses, problem-solving process, progress tracking, and standardization.

Table 5.2 Case Studies PDKA Matrix

PDKA Principles/ Companies	Executive Buy-in	Action Plan	Teams	Consoli-dation of Information	Orga-nizing Frame-work	Identification of Strengths and Weaknesses	Clear Decision-Making Criteria	Prioritization of Strengths and Weaknesses	Assign-ment of Tasks	Problem-Solving Process	Progress Tracking	Periodic Reviews	Standard-ization
BTNI	X			X	X		X					X	
BI Perf. S.			X	X	X	X	X		X			X	
Fed S.S.				X	X	X	X					X	X
Milliken				X	X	X	Informal		X				
NASA				X	X		X					X	
NIQA	X			X			X		X			X	
Pal's S.S.				X	X	X	Informal		X				
PQAP	X			X	X		X		X			X	X
Red River	X			X	X	X	X			X			
Royal Mail				X	X	X							
TVA F&HP				X	X		X	X	X	X		X	
Ulster				X	X	X	Informal		X				
ESQA	X			X	X		X		X			X	X
UWMT	X		X	X			X	X				X	
USARDEC	X			X			X			X	X		

If award-winning companies and organizations, especially those just beginning their quality initiatives, have weaknesses, they are probably failure to choose the most important data, prioritize the strengths and weaknesses to attack first, solve the problem, make sure the solution that was implemented really works, and finally, standardize that solution and incorporate it into other parts of the company. These are the very same weaknesses that led to the development of the PDKA process.

CONCLUSIONS

As we have seen, the PDKA process evolved from the PDCA process. It was an attempt to correct the weaknesses found not in the PDCA process itself, but in the application of organizational information within companies. The failure of companies to integrate and deploy feedback that could help the organization improve, or in other words help the organization to complete the Check and Act steps of the cycle, is problematic. The PDKA process offers a step-by-step methodology to complete the cycle and, as a result, experience success.

As shown in the case studies to be presented, many award-winning companies and the awards programs themselves have benefited from the principles incorporated in the PDKA process, even if they were not aware of the process itself. Even greater improvements are sure to follow if the 12 PDKA principles are structured by and incorporated into the complete PDKA process.

Each turn of the wheel, that is, each cycle of improvements, will surely lead to greater knowledge. The status of World Class is a goal, but it is not a goal that can be kept once it has been reached. The awards are not the end, but the beginning. To maintain the status of World Class requires continuous improvements.

The search for knowledge is a journey. Upon reaching the end, one finds it is not the end but only the beginning.

6

HOW THREE QUALITY AWARD PROGRAMS IMPLEMENT PRINCIPLES OF PDKA—THE PRESIDENT OF THE UNITED STATES QUALITY AWARD, THE UNITED WAY'S EXCELLENCE IN SERVICE QUALITY AWARD, AND THE NORTHERN IRELAND QUALITY AWARD

INTRODUCTION

Two U.S.-based awards programs have actually integrated the logic of the Plan-Do-Knowledge-Act process into their programs: The United Way of America's Excellence in Service Quality Award (ESQA) for the not-for-profit human service sector and the President of the United States Quality Award Program (PQAP) for federal organizations. Likewise, one European quality award program, the Northern Ireland Quality Award (NIQA), has also integrated the logic of the Plan-Do-Knowledge-Act process into its program.

All three awards programs establish performance excellence programs that measure not only financial performance such as cost savings or revenue growth, but also pay equal attention to overall excellence and being a better service provider while doing it with less money. The two U.S.-based awards programs make use of the Malcolm Baldrige National Quality Award (MBNQA) criteria. The Northern Ireland Quality Award is based on the European Foundation for Quality Management (EFQM) Business Excellence Model. This model has nine criteria to assess an organization's progress toward excellence based on the premise that customer satisfaction, people (employee) satisfaction, and impact on society are achieved through leadership driving policy and strategy, people management, resources and processes, leading ultimately to excellence in business results.

The Excellence in Service Quality Award

The United Way of America's Excellence in Service Quality Award (ESQA) recognizes not-for-profit human service sector organizations for performance excellence and service sector leadership improvement. Award recipients need to demonstrate integration throughout their management system and improvement results over a wide range of indicators—customer-related, operational, and financial. The reported results need to address all stakeholders—customers, employees, volunteers, suppliers, partners, and the public.

ESQA has a feedback implementation learning tool that is a guide for not-for-profit human sector organizations to use. The tool is designed as an eight-step process, each step building on the other. If a company follows this step-by-step process, it automatically uses the tool. The tool follows a PDKA logic flow in outlining how not-for-profit organizations should make use of the organization's information.

The President's Quality Award Program

The President's Quality Award Program does not have a PDKA training component, but it does have annual conferences where speakers lecture to potential applicants on the issue of PDKA logic, as well as prepare papers that discuss this subject. The PQAP encourages organizations that have gone through an assessment to actually use the feedback as a learning tool and a guide to making improvements.

The performance excellence criteria for the President of the United States Quality Award Program are closely aligned with the Malcolm Baldrige National Quality Award Criteria (MBNQA), with several modifications to reflect the government environment.

- Terminology is changed selectively throughout the document to reflect the government rather than business environment.
- The point values for several categories and items are slightly different.
- Item 7.2, Overall Financial and Performance Results, is modified to provide for reporting on overall performance results in both financial and non-financial terms.

The close alignment with the MBNQA is designed to foster cooperation and exchange of information between public and private sector organizations, and to ensure that the best practices of high-performing organizations are reflected in the government. The criteria are updated on an annual basis to reflect the best approaches within the private and public sectors to systematically improve performance. The criteria are used extensively throughout the federal sector as a proven framework to guide customer-focused performance improvement efforts.

The Northern Ireland Quality Award

The Northern Ireland Quality Award criteria are very similar to those of the PQAP and ESQA (see Figure 6.1). The model's nine boxes represent the criteria used to assess an organization's progress toward excellence, grouped into "Enablers" and "Results." The Results side of the model is concerned with what the organization has achieved or is achieving; the Enablers describe how those results are being achieved and are therefore better indicators for the future. The maximum number of points available for each criterion is shown, as used for scoring self-assessments and applications for the award. The equivalent percentages indicate the criterion's relative importance to the whole.

To most businesses, large or small, "results" are often the most important aspect. NIQA's model emphasizes that organizations must also consider the underlying processes, methods, and procedures that ultimately lead to the results.

NIQA's award process is designed to recognize excellence in differing sectors (large organizations, small organizations <50, service, public sector). While the criteria (and associated subcriteria) are generic in nature, customized versions of the indicators organizations are required to address are available. An organization that enters the award process is required to demonstrate that through a structured, integrated approach to managing Enablers, it is achieving sustainable results. Significantly, it is essential to demonstrate that the organization understands how the results are achieved.

Given that the criteria, subcriteria, and associated areas to address are generic, there is the opportunity for quality award recipients and indeed

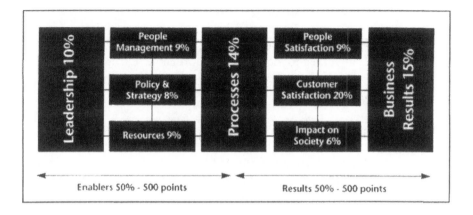

Figure 6.1   NIQA Criteria

other entrants to benchmark against internationally recognized best-practice indicators.

The Feedback from the Three Awards as a Framework

Each of these three awards programs provides guidance for the applicant organizations in the form of feedback reports. These feedback reports are a form of information the organizations can use to help determine opportunities to deploy. The feedback reports also are organized by a performance-based criteria, which is a type of framework. So the feedback reports from the award programs can be used as the base document that all other forms of information in an organization can be integrated into. In other words, the feedback report becomes the organizing framework and catalyst for consolidating all of an organization's information.

HOW THESE PROGRAMS HAVE IMPLEMENTED THE PRINCIPLES OF PDKA

Executive Buy-in

The PQAP process states that, prior to having an assessment performed, an organization's senior leaders need to answer certain questions honestly. They need to answer questions such as: Do you have a plan for where you want the organization to go in the future? Do you have a fact-based, prevention-oriented improvement system? Have you successfully implemented changes consistent with feedback received from prior applications to PQA or other assessment processes? If an organization cannot honestly answer "yes" to these questions, then the organization is probably not using its current information effectively and thus should not apply for more feedback.

ESQA's tool suggests that senior executives must understand the feedback and accept the need to use this feedback for improving the organization. For this reason, they must review the findings from the feedback and make sure that they indeed understand the findings. Once management clearly understands this report and has no need for further clarification, the information contained in the feedback can now be considered potentially actionable.

The NIQA process requires an organization to have a structured approach in place for reviewing performance. This is not merely a review of activity but also a fundamental review of the overall approaches in place. That is not only applicable to strategic and planning processes (i.e., relevance of data inputs to planning process; rationale for objectives and associated targets) and people development issues (i.e., alignment of individual objectives with company objectives), but also to leadership behaviors. The leadership criteria focuses on the extent to which managers at all levels display the attributes of leaders and requires leaders to review the effectiveness of their leadership style (in terms of impact on planning processes, allocation of resources, communication effectiveness, etc.).

Given that award recipients must demonstrate that, trend data must be in place to demonstrate the effectiveness of approaches, and the onus on leaders is not only to understand, but also to act on feedback from internal reviews/assessments.

Consolidation of Information

The PQAP feedback comments provide an objective perspective about the applicant organization within a common criteria-based framework. The PQAP process states that the key and focus of an organization's efforts should be integrating change opportunities into its strategic planning and into business processes and identifying both near- and long-term strategies to drive improvement. An organization can do this by assembling all prior feedback it has collected, then seeking to understand and gain knowledge from all of its feedback. The PQAP states that winning is not the most important aspect of the award; the feedback an organization receives, whether it wins an award or not, is more important than the award itself. The PQAP process acknowledges fully the criticality of integrating and using organizational feedback. It also expresses that organizations can do a better job of getting good assessment feedback if they do a better job of writing their applications so that examiners can better understand the business they are assessing and thus do a better job of providing high-quality feedback. The PQAP process also states that feedback can be enhanced if care is taken to be organized and ready for the examiners' site visit. Again, this will enable the examiners to provide a higher quality, more exact feedback report.

The ESQA learning tool recommends that action plan team members be responsible for consolidating information sources into a feedback summary by integrating the assessment with other existing information sources in their organization. Each piece of information should be considered an opportunity area. The action plan team will then determine the importance of each opportunity area for the organization. The key is to choose only a few areas to improve.

The NIQA feedback report provides an objective, nonconsultative review of an organization's status against the model criteria. Even if organizations do not receive a site visit (not all entrants are site-visited, as a certain point level that organizations have to attain is determined by the judging panel) and do not receive formal recognition within the process, the feedback report represents a critical aspect of the improvement process. While it is true that the quality of the submission document can influence the accuracy and relevance of the feedback report (particularly if organizations do not receive a site visit at which information/written evidence can be validated), there is sufficient rigor and objectives built into the process to ensure that the feedback does provide an objective validation of what the organization acknowledges to be areas for improvement. Ultimately of course, the key determinant of successful utilization of the feedback is the extent to which the areas for improvement are integrated into the planning process and subsequently reviewed to determine levels of resultant improvement.

Decision-Making Criteria

PQAP says to prioritize, since you cannot fix everything at one time. Consider using a matrix with the following criteria areas (similar comments received, comments consistent with your strategic direction, systematic issues that have broad implications for the organization, something important to the customer, something with potential significant impact on improving the business, something your competitors would profit from if they knew your present capability, something you can change and want to change, and something you are willing to commit to improving) and rate them high, medium, or low.

The ESQA tool suggests how to decide on which opportunity areas to proceed. The action plan team selects the opportunities for implementation from among both the strength and weakness areas. Organizations must determine priority based on what they believe are their key business factors. In other words, they need to determine what the important opportunities are.

Given that there are some 32 subcriteria against the 9 main criteria of NIQA it is imperative that organizations prioritize the areas for improvement highlighted. It may be that there are common themes that run through the feedback (i.e., effectiveness of communication mechanisms) and have wide-ranging impact on organizational performance. Organizations may also wish to highlight short-term improvement activity to generate momentum behind overall initiatives. Whatever the rationale, there needs to be a strategic, systematic focus on the identification, management, and review of improvement opportunity. Furthermore, organizations that have a mature approach to managing improvement can also use the scoring process as a valid measure of the process. Repeated self-assessments facilitate a clear understanding of capability that when assessed against the scoring profile allow the organization to determine an accurate score. The more rigorous the assessment process, the more accurate the score as a consequence of the PDKA cycle being in place.

Assignment of Tasks

The ESQA tool recommends organizations to form task teams to develop solutions for the opportunity areas. All selected opportunity areas should become "tasks" that can be factored back into the organization by using a disciplined problem-solving process to attack the areas of opportunity. Task teams are improvement teams whose members are appointed based on their expertise with the problem.

While no prescriptive mechanisms are highlighted as part of the NIQA process, organizations will inevitably adapt a team-based approach to managing improvement activity. While it may not be feasible to action all the improvement areas identified, those that are managed in a structured manner (i.e., appropriate allocation of expertise to team; concrete deliverables and milestone achievements identified; cost–benefit analysis conducted; measures of effectiveness established, etc.) will be implemented.

Periodic Reviews

The ESQA tool recommends that organizations should integrate all selected opportunities into business plans. Once improvement opportunities have been integrated into business plans and their progress tracked, the feedback implementation process is in place. However, management must continue to conduct business plan reviews, which include the status of the tasks and the implementation of the solution for the opportunity areas.

The PQAP further states that it is important for organizations to integrate improvement objectives and strategies into their strategic or business planning process; provide resources consistent with their plans and the importance of their improvement efforts; integrate progress assessments into business plan reviews; make changes where appropriate; formulate action plans consistent with the focus of improvement efforts; integrate thoughts and ideas, and involve those who will make the vision of improvement a reality; and communicate emphasis and involve people in the excitement of improving the business.

A prerequisite of valid, measurable improvement is the extent to which improvements as a consequence of review become integrated with normal working practice. The emphasis on review within the NIQA process reinforces the idea of review of activity as well as a review of overall approach. For example, areas for improvement can be incorporated into the planning process and revised objectives can be mentored on a regular basis, but the true review of effectiveness emphasizes the consideration of the basic rationale behind approaches in place. For example, what *relevant* information is used in the planning process; how is this information assessed for ongoing relevancy.

As another example, many organizations are adapting the ISO 9000 standard as a means to manage their processes. The rationale of review is inherent in the internal audit aspect of the standard criteria. But how often is the ISO 9000 standard itself reviewed for effectiveness in helping meet strategic objectives?

Standardize the Improvements

All three awards programs understand the importance of fixing a problem in one place and then duplicating that fix in other places where the same problem exists. In other words, it is important to standardize solutions throughout an organization's business system. NIQA, as well as ESQA and PQAP, can obviously help identify good practice. Internal benchmarking can be facilitated by acting on feedback from assessments.

CONCLUSIONS

Assessments are not enough. The President's Quality Award Program case study focuses on four steps that outline their overall philosophy to assessment and organizational improvement. The four steps are Plan, Do, Knowledge, and Act. The Excellence in Service Quality Award Program case study explains an organizational learning tool that has eight steps that parallel the PDKA logic. Many awards programs are designed to assess and provide feedback to organizations about what they do well and what they do not do well. The feedback comes in the form of information bites, partial snapshots in time that give guidance and suggestions about the current effectiveness and efficiencies within companies. These pieces of information are exactly that; they may or may not be 100% accurate. If companies combine their feedback report with other sources of information, such as employee surveys, employee focus group comments, customer surveys, supplier surveys, partner feedback, and performance gaps found in key indicators, then they may be able to see what the critical issues really are.

Likewise, the NIQA emphasizes that progress as an organization depends on understanding what influences performance levels. While the assessment process in itself provides a moment-in-time reflection, the overall emphasis is on benefiting from a balanced perspective on performance. There will be an inevitable time-lag between initiating improvement and generating valid measures of progress, but the process reinforces the PDKA logic.

In summary, the PDKA logic which has been integrated into these awards programs allows organizations to be aware of the value of feedback as well as the value of converting the feedback opportunities into knowledge, and then acting to deploy the solutions to these opportunities into the business. The Quality Award feedback report is a good tool because the examined companies are given a CHECK of their businesses, but unless companies ACT upon this information to make meaningful improvements, then the feedback report is of little value. Most of the awards programs are not designed to assist companies in consolidating the feedback information with other organizational information to obtain knowledge or in helping the companies prioritize their intergrated feedback so that they can obtain actual improvements. This assistance may or may not be the role of the quality award programs, but actually taking any kind of feedback and moving it into implementable actions that lead to improved business results is critical to any company. With the PDKA approach, the organizations that participate in each of these awards programs can ensure that they work on only the most important areas of opportunities for the future. As a result of these types of systematic improvements based on implementing solutions to the most critical opportunity areas, companies throughout the world have made significant improvements to their bottom lines.

PROGRAM

The President of the United States Quality Award

PRODUCT/SERVICE:

- Provide assessments for federal organizations in the form of feedback reports
- Award federal organizations as:
  - Presidential Award for Quality Winners
  - Award for Quality Improvement Winners

LOCATION:

Washington, D.C.

KEY PERSONNEL:

Case Study Contributor: Barbara Smith and Don McLeod

KEY USES OF INFORMATION:

Performance Excellence Criteria:
    Leadership
    Strategic Planning

Customer Focus
Information and Analysis
Human Resource Focus
Process Management
Business Results

ORGANIZING FRAMEWORK TO ASSIST ORGANIZATIONS IN CREATING KNOWLEDGE:

The President of the United States Quality Award Program Performance Excellence Criteria

SIGNIFICANT VALUE ADDS AND RESULTS:

The President's Quality Award Program provides a unique opportunity to transform an organization into a more productive, effective, and higher-performing operation. Hundreds of federal organizations have applied to the program. Each year, winners have been recognized through two awards: the Presidential Award for Quality and the Award for Quality Improvement. Applications are reviewed by a team of examiners who are well trained to use the criteria for evaluation and who adhere to clear guidelines regarding confidentiality and conflict of interest. Applicants consistently report that they reap greater benefits from applying to the program than from winning an award. Applicants receive a detailed feedback report, highlighting strengths and improvement areas. The objective feedback from outside experts experienced in quality improvement helps the organization prioritize and target improvement opportunities.

APPLYING PDKA LOGIC TO THE PRESIDENT OF THE UNITED STATES QUALITY AWARD PROGRAM

Getting the most from the President's Quality Award (PQA) Program requires knowledge and understanding of the process and using the best methods to prepare for participation. Making the process work better for everyone depends upon preparing well, executing well, and sharing information and improvement insights. This discussion focuses on some practical guidelines for federal organizations to consider when undertaking such a program. Some of these guidelines may seem obvious, but they are often overlooked, leading to less-than-anticipated satisfaction with the outcomes.

The Excellence in Service Quality Awards program, through its eight-step learning tool, attempts to bring breakthrough improvements to the

United Way of America and to U.S. not-for-profit human service organizations. The President of the United States Quality Award program has offered similar training and presentations to interested federal organizations. Seminars in how to use the President's Quality Award Program (PQAP) feedback are typically offered during its annual awards conference. The intent of these sessions is to help organizations learn how to actually do something productive with their feedback reports.

The President's Quality Award Program and the associated process is intended to

- Recognize federal government organizations that have improved their overall performance and capabilities and demonstrated a sustained trend in providing high-quality products and services, effectively using taxpayers' dollars.
- Promote sharing of the best management techniques, strategies, and results-oriented performance practices among all federal government agencies, as well as state and local governments and the private sector.
- Provide models for other organizations to assess their overall performance in delivering continuous value to customers.
- Provide a systematic, disciplined approach to deal with the dynamics of change by providing a working tool (framework) for conducting assessments, analysis, training, and performance improvement planning.

A commitment to PQAP is a commitment to the achievement of excellence through customer-focused continuous improvement. The PQAP criteria provide a blueprint for superior organizational performance and are used to assess the current standing of an organization. The process to be followed has four steps: (1) documenting the organization's approach to business (the application); (2) evaluating this approach based on an accepted set of criteria (including site visits); (3) developing feedback reports along with PQAP recognition appropriate for the level of attainment; and (4) applying the integrated feedback to promote implementation of improvement. Each of these steps carries with it special requirements and responsibilities for both the applicant and PQA program administration: Office of Personnel Management (OPM) staff, examiners, and judges.

The following steps apply the PDKA logic to The President of the United States Quality Award Program to help federal organizations win in more ways than one. Winning is not the most important aspect of the award; the feedback an organization receives, whether it wins an award or not, is more important than the award itself.

Step One: PLAN

Applying for the President's Quality Award Program—Ready or Not

Before an organization decides to apply for the President's Quality Award Program, there are some basic questions it needs to ask about its current state of readiness. The checklist below will help in this assessment. The organization needs to answer "yes" or "no" to these questions.

- Do you have a plan for where you want the organization to go in the future?
- Are your customers segmented and do you know their requirements?
- Do you know your key processes and have systems in place to drive performance improvement?
- Are the key processes well defined, integrated and aligned to your key business objectives?
- Have you established a systematic approach to your business?
- Do you have a fact-based, prevention-oriented improvement system?
- Have you successfully deployed a quality management system?
- Do you have results data to validate the effectiveness of the approaches taken?
- Do you have data to demonstrate performance as compared to external organizations/companies with established benchmarks and competitive comparisons?
- Can you provide evidence of a consistent positive trend in evaluating performance and initiating cycles of improvement?
- Have you successfully implemented changes consistent with feedback received from prior applications to PQAP or other assessment processes (if appropriate)?
- Can you realistically serve as a role model of quality excellence for others?

Deciding Whether to Perform a Self-Assessment or Have a PQAP Assessment Performed

If the answer to most of these questions is "no," consider conducting a self-assessment using the PQAP criteria instead of applying for the President's Quality Award.

The self-assessment process is similar to submitting an application, except that the organization performs the steps that the PQAP administration would handle; that is, the organization will assess and develop the feedback report rather than submit information to others for their evaluation. Self-assessments can provide all the benefits of the PQAP (except the award), including an independent evaluation if the organization chooses to involve others outside its organization. This is the best option for organizations that are beginning to develop their quality/continuous improvement system. Regardless of which route is chosen, the preparation process will improve the organization.

Step Two: DO

Organizing a PQAP Application or Self-Assessment

There are many ways to organize the development of a PQA application or self-assessment. Each organization, depending on its own uniqueness, must decide the best process to use. As food for thought, consider the following four-phased process that some organizations have used.

The *first phase* is to get everyone who will be helping in the development of the application involved and review the mission and vision, list key external customers (grouped into segments) and suppliers, identify the key processes that develop and deliver the products and services to the customers, and determine the most important measures of effectiveness.

The *second phase* is to set up cross-criteria teams to conduct assessments for customer groups based on customer focus, process management, and business results. The focus here is on the customer requirements by customer group. When these assessments are complete, the information is shared with all involved and serves as the foundation for the third phase.

The *third phase* is to establish teams for the other categories. Some teams will assess support services and suppliers from both a process and results perspective while others will do the same for the categories of Leadership, Information and Analysis, Strategic Planning, and Human Resource Development and Management.

The *fourth phase* is a detailed review of all the categories starting with the Customer, followed by Process Management, Results, Strategic Planning, Human Resource Development and Management, Information and Analysis, and finally, Leadership. This detailed review is designed to identify strengths and areas of improvement; ensure consistency, alignment, and integration for the PQA criteria; prioritize opportunities for improvement based upon importance and impact; identify gaps; and determine and document improvement action plans. This last step, particularly the prioritization of opportunities and development of action plans, is discussed in detail later.

By following this four-phased process (or some other systematic process), an organization will shorten the cycle time for application preparation and enhance the quality of the product. This method provides a greater understanding of the entire assessment process, highlights the linkages between the criteria, and decreases the number of rewrites necessary to eliminate duplication from one category to another.

Preparing the Application

In preparing your application, you need to understand the critical importance of clarity in your presentation. Without it, the process suffers—from everyone's perspective. Clarity is essential in this process because of the time and page limitations involved. Here are some other common ways to improve your application:

- Involve and obtain a commitment from your leadership.
- Use a logical flow when explaining your business, be consistent in content from item to item, and be certain that the links between categories as they relate to your business are explained clearly.
- Write for the examiners. They are your customers.
- Make the application interesting and logical.
- Use facts to provide reasonably supported evidence; limit anecdotal discussions.
- Be responsive to the criteria: do not make examiners search for a response.
- Be clear and specific, and clearly explain the most important aspects of your business.
- Use active voice (e.g., "we implemented the procedure" rather than "the procedure was implemented").
- Minimize the use of acronyms.
- Ensure that all graphics are readable and consistent.
- Use results data to show trends, relationships to comparable high performing organizations.
- Be sure to describe the evaluation and review processes: cite examples.
- Put your own personality into it. Make it fit you: do not copy someone else.
- Provide a central theme or focus associated with your business to link discussions, evidence, and data, and to maintain consistency and continuity throughout.
- Use pictorial displays to clearly describe your processes.
- Seize the opportunity, make the commitment, present reality.

One of the most important steps in preparing an application is developing the Organizational Overview. The overview will set the stage for the details of the seven categories by specifically noting what is most important to the organization and how it conducts its business. Keep the message simple, straightforward, clear, and specific.

While the categories of the President's Quality Award Program have independent requirements, they also weave together to form a tight fabric. Failure to recognize this can lead to a fragmented document and leave the examiners with a perception that the applicant organization does not have a well-integrated business management system. The application process is designed to evaluate an organization based on its systematic approach to business, fact-based decision making, orientation of prevention, full deployment of well-designed approaches, cycles of improvement, and results consistent with approach. The process is also designed to show that these approaches are clear and sustainable over time and compare favorably to the approaches of other organizations that have been noted for their excellence in performance. To move to higher levels of recognition, organizations must demonstrate that they have maturity in each of these areas and clearly communicate this accomplishment.

Preparing for Site Visits

Site visits are used by examiners to *verify* the accuracy of the application, *clarify* uncertain points in the application, and gain *additional perspectives* not included in the application. Examiners review those areas that are most difficult to assess in a written application: deployment, integration, and ownership. Examiner teams will identify site visit issues based on the need to verify or clarify strengths or areas for improvement. Size and complexity of the applicant will influence the visit length, which is usually 4 to 5 days. Eliminating surprises is of primary importance; therefore, both the organization and visiting team need to plan adequately to ensure that the visit is a success. For the site visit to be successful, both the applicant and the examiners must reach a mutual understanding, from the beginning, concerning the purpose and scope of the site visit. Failure to do so can lead to frustration on both sides. Gathered from past experiences of examiners, here are some tips for applicant organizations to remember:

- Prepare, prepare, and prepare.
- Reach a common understanding of the purpose of the visit up front (team leader and applicant).
- Develop specific agendas, but be flexible; these will likely change.
- Provide the team with schedules of events occurring during the visit; however, do not schedule or automatically expect the team to attend any specific event (the team will decide what to attend).

- Be responsive to requests for information: do not require examiners to search for information they need. Examiners should request in advance information that requires special preparation.
- Do not flood the examiners with data and information they do not request.
- Do not try to "do a sales job" on the examiners.
- Make sure senior executives are available for discussions.
- Minimize presentations: they drain time from the visit and try the patience of examiners.
- Prepare the staff for the visit: they should be aware of the what and why of the visit.
- Respond to questions in a specific manner: do not launch into unfocused storytelling.
- Think of the examiners as guests who have a specific mission and a limited amount of time to accomplish it. It is important for them and for the applicant to manage time wisely. Select a good coordinator to ensure that all logistics are addressed.
- Feedback on the application and the site visit will not be provided during the visit.
- Understand that the lack of questions by the examining team about a certain area does not mean it is not highly regarded (or that it is). It means that the team understands the approach and agrees on its value.
- Understand that interviews will be targeted and that many employees will not be interviewed.
- Questions you might expect include: How often? When? How many? May I see examples of ____? May I see a copy of ____?

Remember, a site visit is a form of recognition for the applicant organization. Be responsive to the needs and purpose of the visit. If done well, it can be a very productive experience for all.

Step Three: KNOWLEDGE

Optimizing the Use of Feedback Reports

Without question, the greatest potential long-term benefit of the PQA process is the feedback provided to the applicants. Without this information, organizations lose valuable insights from knowledgeable individuals who wish to help promote improvement. The comments provide a fresh, independent, and objective perspective about the applicant organization within a common criteria-based framework. Why is it, then, that organizations so often fail to recognize the value of this information and how to effectively use the feedback? Perhaps they lack a systematic approach to implementation.

The feedback report you receive will describe certain findings related to the events you presented in your application. While your first reaction to the feedback may generate various levels of emotion, you should look beyond your feelings and focus on the facts to objectively determine what you can learn and use to promote improvement. Just as you organized your application, develop a systematic approach to diagnose your feedback and determine what steps to take and/or changes to make. What you do *not* want to do is disregard your feedback, minimize its value, or disqualify its applicability.

While there are different approaches you can take, the key and focus of your efforts should be integrating change opportunities into your strategic planning and into your business processes and identifying both near- and long-term strategies to drive improvement. To define an approach that works well for you, consider these factors:

- Assemble all feedback you have collected (prior reports, employee surveys, and self-assessments).
- Review your feedback thoroughly—seek to understand and gain knowledge.
- Clarify your understanding and validate the message in the comments.
- Prioritize, since you cannot fix everything at one time. Consider using a matrix with the following topic or criteria areas and rate them high, medium, or low or use other quantifying factors:
  - Similar comments received (prior reports, employee comments/suggestions)
  - Comments consistent with your business/strategic direction
  - Systematic issues that have broad implications for the organization
  - Something important to the customer and customer satisfaction
  - Something with potential significant impact on improving the business
  - Something your competitors would profit from if they knew your present capability
  - Something you can change and want to change (distinguish between near-term and long-term)
  - Something you are willing to commit to improving
  - Others (expand your matrix to cover those areas of significance to your business)
- After determining which elements of your business you will attack first, establish a cross-functional team to develop recommendations

or improvement strategies or both. Be sure to consider a systems integration perspective.

■ Integrate improvement objectives and strategies into your strategic or business planning process.

■ Communicate your emphasis and involve people in the excitement of improving the business.

■ Formulate action plans consistent with the focus of your improvement efforts. Integrate thoughts and ideas, and involve those who will make the vision of improvement a reality.

■ Provide resources consistent with your plans and the importance of your improvement efforts.

■ Integrate progress assessments into business plan reviews. Make changes where appropriate.

Step Four: ACT

Keep in mind that the feedback report from PQAP can trail the development of an application by as much as 6 to 12 months, depending on when you begin. There is, however, opportunity to begin assessing your improvement opportunities early in the process. As you develop your application, it will become evident which areas need greater attention. Do not lose this opportunity. Begin right away to assess your understanding and promote improvement. If you do nothing else, perform a self-assessment of your application once it is completed. It will give you some insights for action as you await your feedback. It will also give you a better awareness and understanding of your external feedback when you receive it. Before you distribute the feedback report to your senior managers, get them together and ask what they think the report might contain. This challenges them to anticipate what they might learn from the report, and it can lead to some interesting discussions.

Here is the critical question: What are you going to do with this information? If you do little or nothing, you have wasted your organization's time and energy. You may have also caused some serious damage to the personal credibility of those involved in the PQAP process and the credibility of senior managers and your business improvement efforts. Much can be gained or lost.

Participation in the President's Quality Award Program can be a rewarding experience in many ways. The recognition that award winners receive is the icing on the cake for the hard work and dedication the organization has devoted to its continuous improvement efforts.

Knowing when to apply is an important consideration. As with any competition, prepare well to compete. Self-assessments can be a valuable tool in this effort. If you enter the competition, you must be able to demonstrate to others how much you have achieved. Perhaps the best measure of the maturity and seriousness of your organization is how it uses feedback to drive continuous improvement. If you seek recognition for recognition's sake, you have truly missed the real value of participation! Win or lose, you must implement solutions in problem areas and drive improvement.

PROGRAM

The United Way of America's Excellence in Service Quality Awards

PRODUCT/SERVICE:

- Provide assessments for not-for-profit human sector organizations in the form of feedback reports
- Award not-for-profit human sector organizations as
  - Platinum Award Winners
  - Gold Award Winners
  - Silver Award Winners
  - Bronze Award Winners

LOCATION:

Alexandria, Virginia

KEY PERSONNEL:

Case Study Contributors: Debra Gittens and William Phillips

KEY USES OF INFORMATION:

Award Criteria:
Leadership
Strategic Planning
Customer and Service Sector Focus
Information and Analysis
Human Resource Development and Management
Process Management
Business Results

ORGANIZING FRAMEWORK TO ASSIST ORGANIZATIONS IN CREATING KNOWLEDGE:

The Excellence in Service Quality Award Criteria

SIGNIFICANT VALUE ADDS AND RESULTS:

The award recognizes organizations for performance excellence and service sector leadership improvement. Award recipients need to demonstrate integration throughout their management system and improvement results over a wide range of indicators: customer related, operational, and financial. The award assessment is tailored to an organization through a focus on factors important to the organization, strategy, and competitive success. These factors are defined by the organization in the Organization Overview, as well as the Strategic Planning Category. Organizations receive a detailed, confidential feedback report, which outlines by category the organization strengths and areas for improvement. Use of the award criteria as the foundation of an ongoing cycle of assessment and improvement leads to the integration and alignment of numerous activities often only loosely connected. The assessment provides an effective means to measure progress and to focus everyone in the organization on the same goals.

THE UNITED WAY OF AMERICA'S EXCELLENCE IN SERVICE QUALITY AWARDS PROGRAM

The United Way of America has created a learning tool that has the potential to support its Excellence in Service Quality Awards (ESQA) program, as well as having the potential for member organizations to use in their self-assessment efforts. This learning tool was developed in the mid-1990s and was shared at various United Way of America forums, conferences, and events. The learning tool (called the *feedback implementation process*) was not mandated and was left optional for the United Way member organizations. The feedback implementation process has varying degrees of use of the learning tool among the member organizations. The United Way of Middle Tennessee has used the tool to further implement improvements in its operations. Other member organizations are using the tool as a guide or as another source of helpful information, and some member organizations are not using the tool at all.

The varying degrees of using the tool is the result of differences in organizational maturities and their varying applications of the business excellence criteria. For example, some member organizations are more familiar with, and mature in, their application of the Excellence in Service Quality Awards criteria; those that are more mature in their quality journey are more likely to understand and apply the feedback implementation process. All of the member organizations have a common goal to raise a certain level of funds in their campaigns while striving to meet or exceed customer expectations surrounding their fund-raising campaigns and the implementation of the programs throughout the year. This learning tool is meant to be one of many in the United Way of America's Excellence in Service Quality Award's tool box to assist the member organizations in achieving higher levels of customer satisfaction. The United Way of America originally introduced this learning tool to provide its member organizations yet another avenue by which to improve efficiency and effectiveness for its customers, employees, and volunteers.

What Led to the Creation of This Learning Tool?

Gauging an organization's readiness for or progress toward a formal culture of continuous improvement requires both an initial as well as periodic organization-wide assessments. As each resulting improvement plan becomes more and more detailed, these assessment tools will evolve in specificity and rigor. As an organization becomes more committed to and involved in developing a quality system, it will often seek the added benefit of external assessments as a further means to validate its progress and improvement opportunities. The Malcolm Baldrige National Quality Award (MBNQA) and United Way's ESQA are examples of such formal assessments in the private and nonprofit sectors, respectively. Experience has shown that even organizations that have invested significant time and resources in an external assessment do not always know how to make the best use of this feedback in their next round of improvement planning. This training tool is dedicated to helping organizations transfer assessment information, internal or external, into actionable organizational knowledge.

What Assumptions Surround This Learning Tool?

- It is for organizations that have begun their improvement process.
- It is for organizations that understand the need for periodic assessments.
- It can be used by all interested organizations.
- It is designed to accommodate organizations along a continuum that ranges from a fairly preliminary assessment based largely on perceptions to formal assessments using external examiners.
- It will add more value over time.
- It will accommodate organizations of varying size and sophistication and may suggest shortening or deleting one or more steps in the process.

What Kind of Learning Tool is the Feedback Implementation Process?

The feedback implementation process is a knowledge-based process that takes information from feedback reports and other information sources and develops a strategy for implementing areas of opportunity into the core of the business. The strategy expands the Plan-Do-Check-Act logic into a knowledge-based implementation process that couples assessment information with theory in the form of the conceptual guidelines of a national assessment/awards process—such as the Malcolm Baldrige National Quality Award in the private sector and the Excellence in Service Quality Award (ESQA) criteria for the nonprofits—to create organizational knowledge.

What Value Can This Learning Tool (The Feedback Implementation Process) Add to an Organization?

Becoming a world-class organization requires converting information into actionable knowledge. Learning to implement solutions to problems is a critical step for any organization, because it moves organizations toward world-class status. That status equates to improving customer satisfaction and reducing costs, which ultimately contribute to growth. If an organization in the private sector had invested in each publicly owned MBNQA award winner, the organization's investment would have approximately doubled, according to industry studies. While as yet undocumented, the potential for increased revenues for service in the nonprofit sector should mirror these financial gains.

What Is the Logic of the Feedback Implementation Process?

The process of improving an organization can be a simple one, but most organizations do not have the discipline necessary to implement their improvement opportunities. Internally, organizations are proficient at planning and gathering information about themselves and are becoming experienced at doing periodic self-assessments. In other words, they are good at problem-seeking but are not as effective at addressing problems that have been identified.

Many organizations have not figured out how to take the information from formal and comprehensive organization-wide assessments or other sources, such as employee surveys or customer feedback, and use it to improve the business. Businesses need to convert information in the form of feedback to knowledge, by *integrating* the solutions to these problems into their organizational processes.

The United Way of America's feedback implementation process is a knowledge-based process that converts all identified opportunities into knowledge, prioritizes the knowledge before proceeding to develop action plans, then acts to implement the solutions to these opportunities into the organization. With this approach, the organization selects only the most important areas of opportunity for the next year. The areas of opportunity should be linked to a problem-solving process and to a business-planning process. The feedback implementation process begins once an organization completes an assessment at any level, on any point on the continuum. This process can be used at any point on the quality journey, regardless of the sophistication of the assessment. Figure 1 is a flowchart of steps 1 through 8 of the feedback implementation process.

How Do Organizations Use This Feedback Implementation Tool?

The tool is designed as an eight-step process. Each step builds on the other. If a company follows this step-by-step process, it automatically uses the tool. The learning tool was designed by the United Way of America with consultation from this book's author.

Step 1: Make sure that senior executives clearly understand the feedback derived from the assessment

The feedback from any serious assessment addresses both pluses and minuses within an organization. Such feedback should be digested and taken as constructive: rationalizing such comments should be avoided; they should be accepted as they are. Senior executives must understand

THE FEEDBACK IMPLEMENTATION PROCESS

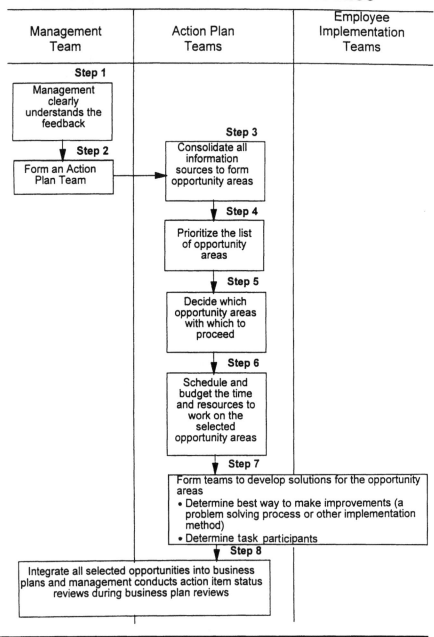

Figure 1   Feedback Implementation Process.

the feedback and accept the need to use this feedback for improving the organization. For this reason, they must review the findings from the feedback and make sure that they indeed understand the findings. Once management clearly understands this report and has no need for further clarification, the information contained in the feedback can be considered potentially actionable.

Step 2: Form an action plan team

The action plan team is composed of a small group of people who will have the authority and accept the responsibility to follow the feedback implementation process all the way through. Senior executives hand the feedback to the action plan team to develop detailed action plans based on the feedback. To do this, the action plan team members first review the feedback to make sure that they, too, fully understand the content.

Step 3: Consolidate all information

Action plan team members are responsible for consolidating information sources into a feedback summary by integrating the assessment with other existing information sources in their organization, such as employee surveys, customer feedback, past assessments, critical organization indicators, and other sources as appropriate. They integrate this critical information by putting it into a matrix that uses seven categories, traditionally seen as comprising a system of continuous improvement—such as the Malcolm Baldrige Award, Excellence in Service Quality Award, and State Awards—as the organizing framework.

Each piece of information should be considered an opportunity area. The action plan team will then determine the importance of each opportunity area for the organization. The team will need to construct two matrixes, one for weaknesses and one for strengths. The team members' goal is to determine those weaknesses in their organization that if improved would most dramatically improve the organization.

Likewise, the action plan team tries to identify those strengths that if improved would move the organization into the world class. The feedback implementation process emphasizes the need to improve strengths as well as weaknesses to become truly leading edge and to stay leading edge. The next step, then, is to develop criteria by which to measure priority; that is, to decide which areas of strength and which areas of weakness will be improved.

Step 4: Prioritize the list of opportunity areas

The action plan team must then prioritize all opportunity areas. First, the team members stratify these data to eliminate duplication. Once stratification has occurred, they must determine the priority of these opportunity areas. They develop criteria by which to measure priority. For each opportunity area, these criteria are quanitified in terms of low correlation, moderate correlation, or high correlation. The scores are totaled and priorities are set by number. The highest point getters and the very lowest point getters suggest the prime candidates to work on. In other words, an organization should work on the few areas of weakness that will improve it the most and work on the few key strengths that if enhanced will turn it into a role model.

Step 5: Decide with which opportunity areas to proceed

The action plan team must now select the opportunities for implementation from among both the strength and weakness areas. The selection scale has been linked with formal Malcolm Baldrige and ESQA assessment criteria, so any strength or weakness area totaling a score equal to or above the cut-off point should be considered as a high leveraged item. However, each organization may opt to decide its own cut-off points. One important thing to remember is that organizations need to work on the higher scoring areas of opportunity, not on everything simultaneously.

At this stage in the process, the information from feedback sources has been integrated with theory to create new organizational knowledge. In essence, this new organizational knowledge has been created with the linking of the matrixes with the seven categories of the Excellence in Service Quality Award criteria.

Step 6: Schedule and budget the time and resources to work on the selected opportunity areas

Next, the action plan team members use the list of prioritized improvement opportunities to determine how many can feasibly be implemented within given time and dollar constraints. The team schedules and budgets the selected opportunity areas and performs a simple cost/benefit analysis, where appropriate, for management. The opportunity areas all have a projected scheduled completion and are tracked against the established schedule. Presentation and review dates are also noted on the schedule. For more involved tasks, the team may also estimate potential costs to implement a solution. Once the proposed actions and any cost/benefit analysis are reviewed by senior-level managers, these tasks are factored into the organization's overall improvement plan and task teams are assigned to deploy improvements.

Step 7: Form task teams to develop solutions for the opportunity areas

One of the most successful techniques for developing solutions for opportunity areas is the development of task teams. Task teams are improvement teams whose members are appointed based on their expertise with the problem. Task teams should be encouraged to use a disciplined problem-solving process to attack the areas of opportunity.

Step 8: Integrate all selected opportunities into business plans

Once improvement opportunities have been integrated into business plans and their progress tracked, the feedback implementation process has been successfully completed. However, management must continue to conduct business plan reviews, which include the status of the tasks and the implementation of the solution for the opportunity areas.

Summary

The United Way of America has recognized that if its member organizations really want to improve their growth and performance, they cannot simply assess their organizations and hope for the best. They must convert information into knowledge and prioritize and implement their findings. They must use the critical information about their organizations and act on newly obtained knowledge by implementing solutions and tracking results so that their organizations will not only grow, but also flourish. However, the United Way of America has also realized that this learning tool (the feedback implementation process) cannot be forced upon the member organization. Instead the tool must be introduced and be available for organizations who are ready for it and willing to implement. This is how the tool will succeed over time.

PROGRAM

The Northern Ireland Quality Award, Europe

PRODUCT/SERVICE:

Provide assessments for organizations who enter the award in the following categories: large, public service, small—across all sectors. Provide recognition to "Excellent" organizations (as measured against criteria).

LOCATION:

Belfast, Northern Ireland

KEY PERSONNEL:

Case Study Contributor: George Wilson and Robert Barbour. Please note that some information was sourced from *Managing Quality*, authors Desmond Bell, Philip McBride, and George Wilson [who provided information for this documentation]; publishers Butterworth-Heinemann.

KEY USES OF INFORMATION:

The Business Excellence Model
Leadership
People Management
Policy and Strategy
Resources
Processes
People Satisfaction
Customer Satisfaction
Impact on Society
Business Results

ORGANIZING FRAMEWORK TO ASSIST ORGANIZATIONS IN CREATING KNOWLEDGE:

The European Foundation for Quality Management Business Excellence Model

SIGNIFICANT VALUE ADDS AND RESULTS:

The Northern Ireland Quality Award provides a framework against which progress and achievement can be mapped. It is not a standard but rather a model for self-assessment that provides a baseline for future measurement. Properly conducted, the quality assessment process produces a benchmark against which an organization's progress can be measured while implementing a quality-improvement initiative. It enables the organization to clearly identify its strengths and the areas in which improvements can be made using each criterion model.

The self-assessment process is more powerful than the traditional quality audit as it encompasses the entire activities of the organization rather than the quality system. The adoption of a strategy that includes self-assessment offers significant advantages to companies who wish to develop beyond the systems approach.

The challenge of making an application for the award will bring your organization substantial benefits. The objective consideration of the key features of an organization's strategy and the development of soundly based improvement plans is the most valuable aspect of adopting the self-assessment approach, which is the first step in any application for the award.

Applying for the award will

■ Sharpen the focus of your organization and its improvement activities

- Foster teamwork—working to tight deadlines provides people with a clear aim and adds an exciting challenge to life at work
- Heighten awareness of business excellence by involving many people in the application
- Create a succinct description (the application document) of the organization in terms of its activities, methods of operation, and results. This will also be valuable in promotion, training, and communication contexts.

A further significant benefit is the feedback report, which is prepared by a team of trained, independent assessors made up of senior managers and quality professionals from throughout Northern Ireland. The report provides a list of strengths and areas of improvement for each of the criteria addressed in the application and includes an overall scoring profile.

There is great value in this process of self-assessment and, even if your organization has little chance of winning an award, the added impetus of going through a complete award cycle enables you to assess your organization's strengths and level of maturity against each category within the Excellence Model.

There is considerable status attached to winning an award. The high-profile promotion of an award winner, together with the opportunity to use the award logo, confirms the organization's position as one of the most successful in the United Kingdom.

Recognition will also be demonstrated by the number of visits requested by other interested bodies and requests for speaking engagements where the winners share their best achievements and are able to promote themselves as organizations of established excellence.

In addition to enjoying the intrinsic benefits gained from implementing a Business Excellence program, winners of the award can also expect to benefit from the emergence of new customers, new business opportunities, and improved customer perceptions of goods and services offered.

THE NORTHERN IRELAND QUALITY AWARD PROGRAM

Sharing Best Practice

In the year following presentation of the award, winners are expected to share their experiences at conferences and seminars organized by the Northern Ireland Quality Center, offering a platform for the promotion of their status as leaders in industry and/or the public sector. Winners will be encouraged to grant the Quality Center the right to use specified material for research purposes. The Quality Center would also anticipate permission to market a nonconfidential version of the winner's submission document.

Assessment

An application is considered by a team of between three and eight assessors (depending upon the applicant's size), all of whom have undergone the same training course to ensure a high level of consistency. Assessors are drawn mainly from the ranks of experienced practicing managers from NI organizations and include some academics and quality professionals.

The organization's application will be assessed for "strengths" and "areas for improvement" and scored on a scale from 0 to 1000 points, using the Business Excellence Model. Having assessed individually, the assessors meet as a team to reach a consensus score for the application. From this initial assessment, the Award Jury, composed of distinguished senior managers and academics, will decide which applicants should be visited. Applicants scoring in excess of an appropriate level receive a site visit from the team to check the validity of the application and to clarify issues of understanding. A final report on the application is then prepared.

After the jury's final review and decision, any awards presented will be to those organizations that demonstrate the highest standards of Business Excellence.

SELF-ASSESSMENT

The technique of self-assessment is essential for any organization wishing to develop and monitor its quality culture. This kind of systematic review and measurement of an organization's operations is one of the most important management activities of any Business Excellence system.

Self-assessment allows an organization to identify clearly strengths and areas for improvement by focusing on the relationships between its people, processes, and results. It should be a regular activity and is an essential prerequisite for an award submission.

Self-assessment and some different approaches to initiating it are described in Quality Centre's literature.

ASSESSING ENABLER CRITERIA

Information is required on

■ How the organization approaches each criterion. Each criterion covers a range of specific questions, and an organization should provide concise and factual information about each of these.

- The extent to which the approach has been deployed—vertically through all levels of the organization, and horizontally through all areas and activities. Ideally, numerical quantification should be provided.

ASSESSING RESULTS CRITERIA

Information is required on

- What the organization is achieving with respect to each criterion. The organization should provide concise and factual information about each of these parts.
- The key parameters your organization uses to measure results and achievements. For each parameter, trends of data are required that cover up to 3 to 5 years. The trends should highlight
  - the organization's actual performance
  - the organization's own targets
  - wherever possible, the performance of competitors
  - the performance of "best in class" organizations
- The rationale behind parameters presented and how they cover the range of your organization's activities. The scope of results is an important consideration for the assessors.
- For each of the results criteria, evidence is required of its relative importance. In the case of Business Results, data may be presented in the form of an index rather than absolute terms to avoid disclosing sensitive information.

To allow comparisons to be made conveniently, it is helpful when the applicant provides a single chart, for each key parameter, showing trends. A brief commentary that demonstrates an understanding of significant features of the data presented is also desirable.

RELATIONSHIP BETWEEN THE CRITERIA

The Business Excellence Model provides a framework for organizations to apply self-assessment and to improve. Assessors will be looking for consistency in the information presented across and between the criteria of the model.

The full power of the model is derived from an understanding of the relationships between criteria. At a basic level, if a process is said to be key in an Enabler criterion, then results to the performance of this process should appear in one of the Results criteria.

While all nine criteria in the model are linked, some relationships are particularly clear; for example,

- People management (criterion 3) and people satisfaction (criterion 7)
- Criterion 4 and 5 (the key processes of the organization) and business results (criterion 9)

Linkages can be expressed between policy and strategy (criterion 2) and the Results criteria. There could also be linkages between policy and strategy and some of the comparisons presented in Results criteria.

For example, if the strategy is to achieve "global leadership," the organization should be seeking global comparisons against which to judge performance. A lesser objective would justify the selection of less ambitious comparisons.

Use of the model as a driver of improvement is also important. It is reasonable to expect a connection between results achieved (and presented in the Results criteria) and actions to improve performance in the Enablers criteria. Comparisons of results with internal targets and competitors should provoke an analysis of the issues driving customer satisfaction and loyalty leading to modifications of policy and strategy (criterion 2) and resulting in plans in the Enablers criteria to achieve improvements.

GENERAL NOTES ON SITE VISITS

The purpose of the site visit is to

- Verify that the application fairly reflects practice within the organization
- Clarify parts of the application that might be unclear
- Allow some new information to be brought forward according to the following rules:
  1. New initiatives started after submission date: These should not be taken account of by the assessors.
  2. Existing activities ongoing before the submission date but not mentioned in the applicant document: If these are brought to light during the visit, the assessor team may consider the additional information and make an appropriate adjustment to the score of the subcriterion concerned.
  3. Continuing data: The applicant may bring forward the latest points on trends of results already presented in the application document. If this is done, then in fairness new points (good and bad) should be brought forward.

The assessor team may consider the additional information and make an appropriate adjustment to the score of the criterion concerned.

- To sense the atmosphere/working environment to test whether or not the company is a role model organization
- To re-score the application based on the visit findings

7

HOW THREE WORLD-CLASS PRIVATE MANUFACTURING COMPANIES IMPLEMENT PRINCIPLES OF PDKA— MILLIKEN DENMARK, PAL'S SUDDEN SERVICE, AND ULSTER CARPET MILLS

INTRODUCTION

The companies that provided these case studies—Milliken Denmark, Pal's Sudden Service, and Ulster Carpet Mills—vary in size and the products and/or services they produce. However, the success of all three case study companies is evidenced by their results.

Milliken Denmark Ltd.

Located in Mørke, Djursland, Milliken Denmark Ltd. produces washable rental mats, mops, soap dispensers, hand soap, scent dispensers, and assorted perfumes. Milliken Denmark is the European headquarters for Milliken's Dust Control Organization. The firm has over 160 employees, of whom about 11 reside abroad. Approximately 100 employees work in production; the rest work in administration and service.

Milliken Denmark has won the European Quality Award and the European Prize, and the parent company has won the Malcolm Baldrige National Quality Award.

Pal's Sudden Service

Pal's is a small, privately owned, quick-service restaurant (QSR) company competing directly with major internationally recognized chains to offer food and beverages ready for customer consumption. Unlike other businesses, a restaurant consists of the combination of manufacturing, service, and retail components all rolled into a single business operation. The manufacturing plant feeds raw materials into scheduled processing, assembly, and packaging steps to produce food items ready for consumption. The service component receives and fills customized orders, and the retail component provides a menu of items for direct sale to buyers.

The success of Pal's Sudden Service is evidenced through a review in the performance areas for its key business drivers (quality, service, cleanliness, value, people, variety, and speed), which have been carefully linked to customer requirements. Pal's consistently wins head-to-head competitions in its market against much larger national restaurant chain outlets. As the number of customers doing business at Pal's increases, its financial performance (sales volume and profit) and market performance (market share) also improve. Pal's has won the Tennessee Quality Award twice and is a recent applicant for the Malcolm Baldrige National Quality Award.

Ulster Carpet Mills Ltd.

Ulster Carpet Mills Ltd.'s primary products are woven Axminster and Wilton carpets. Ulster Carpets has about 1200 employees located in Northern Ireland and South Africa.

Ulster Carpets has seen exports increase 700%, output per employee rise 60%, and sales increase more than 130%. Ulster Carpets has also been winner of the Northern Ireland Quality Award, the United Kingdom Quality Award for Business Excellence, and the United Kingdom Safety Award.

HOW THESE COMPANIES HAVE IMPLEMENTED THE PRINCIPLES OF PDKA

The companies represented in these three case studies have distinctly different product lines; however, all three companies use the broad logic of the PDKA process. While none of the three uses all 12 PDKA principles, they all use some form of gathering and integrating information, and they all use information to improve their businesses. Each of the case studies demonstrates use of variations of the principles of PDKA, and the variations typically occur within the five core principles (Consolidation of Information, Organizing Framework, Identification of Strengths and Weaknesses, Decision-Making Criteria by Which to Measure Priority, and Assigning Selected Opportunity Areas as Tasks).

Consolidation of Information

Milliken Denmark gathers a variety of information types and generally uses the European Quality Award Model to help integrate its activities. Milliken Denmark starts with measurements from a few critical sources. These measurements are then used as facts or information which are further used in making decisions about the business. Milliken uses a customer satisfaction survey, an employee suggestion system, a correction report to achieve zero defects, and other sources such as employee safety and keeping delivery dates.

1. "The Company Barometer," the annual customer satisfaction survey, compares Milliken's customers' satisfaction level with that of its biggest competitors. This is the most important statistic for the company. A sample of customers are interviewed by a bureau on a wide range of issues connected with how they rate collaboration with Milliken Denmark and its competitors. If the survey shows that Milliken's competitors have come out on top, or if Milliken has achieved a lower score than in previous years, then corrective measures are taken. This measurement is regarded as the most important measurement of all.
2. Milliken Denmark has a formal employee suggestion process, the Opportunity For Improvements (OFI). The system deals with more than 4,000 opportunities for improvement per year. The OFI system has proven to be an excellent vehicle for transmitting employees' ideas. All approved suggestions are rewarded with a small prize, the size of which has only a minor relation to the importance of the suggestion for the company. Basically, Milliken believes that the OFI system is in the interest of all employees, ensuring a strong firm, giving them greater job security and the chance to earn more.
3. The "correction report" implements a zero-defect philosophy. The purpose of the "correction report" is to describe defects systematically and ensure that their causes are eliminated. More than 1,000 correction reports have been written, each of which has eliminated at least one source of error.

4. Key focus areas, such as employee safety and keeping delivery dates, are also used to gather facts.

Pal's Sudden Service has carefully designed a system for the collection and analysis of information and data that feeds and interacts with each component of its Business Excellence Process. The Pal's Management Information System is also used to guide the selection, management, and effective use of information and data to support key operational processes, action plans, and the performance management systems. The information and data collected at Pal's are used to produce a Balanced Scorecard of Core Performance Measures and to support the management and continual improvement of overall company and operational excellence. The balanced scorecard links in-house performance measures and data charts for each key business driver with the influencing key operational processes and strategic action plans. In addition, measures are provided to assess the impact of performance for corresponding business support processes and supplier processes. Pal's obtains strategic information from two primary sources:

1. Customer and market data from a variety of sources are gathered to interpret customer needs and requirements, to project market trends, to establish new strategies that will delight customers and sustain competitive advantages, and to check the impact of previous strategies and plans on customer satisfaction.
2. Competitor information from Pal's benchmarking process is used extensively to identify market and industry trends, industry and competitor capabilities and best business practices, competitor strategies, potential competitor reactions to their strategies, and promotional and technological improvement opportunities. Pal's uses a systematic benchmarking process to determine best-of-class in practices and performance and to set stretch goals to reach and exceed best-of-class performance levels. The benchmarking process is part of the Business Excellence Process and feeds comparative data and best business practices to the Management Information System.

Ulster Carpet Mills bases policy and strategy decisions on information that is relevant and comprehensive. The key process for taking all of its organizational information and converting it into business results has six broad steps: (1) planning/forward development plan (scope, comparatives and data); (2) pilot scheme (training and process owner identified); (3) analysis (performance gap and future levels); (4) integration (internal customer–supplier assessment); (5) implementation (action plan, team playing, company brief); and (6) review.

Step 4, integration, is the key step in the consolidation of information. The analysis and the integration steps give Ulster Carpets key knowledge upon which to base implementation decisions. Six key types of information are gathered and used:

1. Customer feedback through retail customer phone and postal surveys, face-to-face formal interviews, trade exhibitions and shows, and customer facing groups within Ulster Carpets. Project teams were set up to advise on how to effectively implement some customer feedback requirement.
2. Feedback from suppliers through monthly review meetings with their transport supplier, visits to the factory, supplier training days at Ulster Carpet Mills, supply contract review meetings, visits to suppliers, new technology and new machine installations.
3. Employee feedback through a variety of avenues, such as employee opinion surveys and project team presentations.
4. Benchmarking studies.
5. Competitor analysis.
6. Information on social issues, including energy use, prevention of pollution, regional and national wages, and community requests for visits, for comparative purposes.

An Organizing Framework

Milliken Denmark uses the organizing framework of the European Quality Award Model. The self-assessment framework of the EQA model greatly helped Milliken to obtain a much better overview of all its quality-related activities, including the management process. One of the key lessons the company management has learned from using the European model is the importance of following the Plan-Do-Check-Act cycle for continuous improvement.

Pal's Sudden Service uses a continual improvement process, featuring a Plan-Do-Study-Act cycle (PDSA), to define standards of excellence, maintain a focus on customers and other stakeholders, and identify, plan, and execute improvement projects. Pal's benchmarking process complements the continual improvement process by guiding Pal's efforts to make meaningful competitive comparisons, to identify and adopt best practices needed to achieve high performance expectations, and to achieve defined directions for the future. It also helps pinpoint the primary areas for organizational and individual opportunities for learning and innovation.

The Malcolm Baldrige criteria is used in conjunction with the continual improvement process as a self-assessment tool driving Pal's toward overall excellence, accelerating implementation of the business vision, and maintaining an emphasis on Pal's processes for delighting the customer.

Ulster Carpet Mills' organizing framework is the company-wide self-assessment using the European Business Excellence Model, which is based on the Malcolm Baldrige Award model.

Identification of Both Strengths and Weaknesses

All three of the case study companies focus on making improvements to areas of weakness; the case studies do not focus significant attention on explaining how they improve their strengths to even higher levels of effectiveness. Focusing equally on improving strengths as well as further improving weaknesses is a subtle part of the PDKA logic that can assist companies in moving further into world-class standing.

Clear Decision-Making Criteria

The three case study companies do not explain a formal quantifiable criteria for making decisions about which opportunity areas to attack first. Milliken Denmark, however, has seven business fundamentals that give guidance as to a logic or way of focusing on certain activities in their business. The three companies clearly have informal methods for prioritizing which initiatives to push forward at any given time and which initiatives to hold for another time or to cancel all together. The PDKA logic states that a formal decision-making criteria can assist a company in determining priority of initiatives that are consistent with the company's values and goals and factor into the company's timetable for needs. Without a clear decision-making criteria, companies risk perhaps working on the wrong things at the wrong times, or even working on conflicting initiatives. A clear decision-making criteria further allows discipline to be built into the process of driving improvements.

Assignment of Tasks

Milliken Denmark uses teams to make improvements. The company assigns selected opportunity areas as tasks. When Milliken has a problem area, management establishes an Error Cause Removal (ECR) team of four to six employees to address the problem. Many employees outside the team are often affected by the problem, or they are aware of it, and they often have opinions on what the cause is and how it can be solved, regardless of whether they are members of the team.

Pal's Sudden Service shares information and organizational learning, which is facilitated by its cross-functional improvement team assignments, where process team members from each store participate in assessment, benchmarking, learning, and improvement projects. Ongoing teams for the seven Malcolm Baldrige categories and continual improvement include members from each store who work on company-wide improvement opportunities and implement common learning and standardized methods.

Ulster Carpet Mills' people involvement in problem-solving was partially led and partially driven in the early years. Over time, involvement has become self-generating as barriers have been broken and fear eroded. They now have around 300 teams in action per year covering all aspects of the business. A key problem-solving aid has been the use of cause-and-effect diagram boards. Each shows the large fishbone diagram with white cards detailing the problem under investigation placed at the "head of the fish." Suggestions of possible causes are then placed around each "rib of the fish," and ideas for solutions are positioned opposite each probable cause.

CONCLUSIONS

Each of the case study companies has excellent information sources and uses these information sources in a comprehensive way that leads to a strategic business knowledge. They achieve this knowledge in slightly different ways. Each company uses some method by which to organize its thinking and its information. Milliken Denmark uses the European Quality Award model. Pal's Sudden Service uses a continual improvement process in conjunction with the Malcolm Baldrige criteria. Ulster Carpet Mills' organizing framework is the company-wide self-assessment using the European Business Excellence model. All three case study companies use teams of people to help implement their opportunities.

COMPANY

Milliken Denmark

PRODUCT/SERVICE:

Washable rental mats, mops, soap dispensers, hand soap, scent dispensers and assorted perfumes

LOCATION:

Mørke, Djursland, Denmark

NUMBER OF EMPLOYEES:

160 employees

KEY PERSONNEL:

Cast Study Contributor: Geoff Carter of the European Foundation for Quality Management, and the European Commission

KEY USES OF INFORMATION:

■ Measurement Types (employee safety, timely delivery, customer satisfaction and failure costs, employee satisfaction, credit memos,

employee education, wasted materials, composition of inventories, and the number of complaints about goods and services)
■ Annual Customer Satisfaction Survey ("Company Barometer")
■ Employee Suggestion System
■ Control Charts
■ The European Quality Award

ORGANIZING FRAMEWORK TO CREATE KNOWLEDGE:

The European Quality Award Model

AWARDS AND SIGNIFICANT RESULTS:

■ Winner European Quality Award
■ Winner European Quality Prize
■ Winner Malcolm Baldrige Quality Award (parent company)

INTRODUCTION AND BUSINESS OVERVIEW

Milliken, as a world-wide organization, has been involved in the quality process since 1981, when the first step was taken to incorporate quality in all aspects of business. This effort has developed over the years, inspired by many scholars, benchmarking, and internal ideas and has evolved to where it stands today—still on the road toward the final goal, Total Quality Management. Although Milliken has the goal to aim for, it knows that what is important is continuous, constant improvement.

To this end, Milliken has, as an organization, worked closely with a number of institutions. Milliken Denmark takes pleasure and pride in working with the Aarhus School of Business in Aarhus and more specifically, with Professor Jens J. Dahlgaard and Professor Kai Kristensen. They have attended many of the organization's internal and external seminars, case studies, and theme days and have also, on occasion, arranged for students to study special subjects within the company.

At the same time Milliken has benefited from the very valuable feedback and inspiration from the school and, specifically, from Professors Dahlgaard and Kristensen, who have helped in Milliken's search for the road toward Total Quality Management.

Context

Milliken Denmark Ltd., located in Mørke, Djursland, is a 28-year-old producer and seller of washable rental mats, mops, etc., as well as soap dispensers, hand soap, scent dispensers and assorted perfumes. Milliken Denmark is the European Headquarters for Milliken's Dust Control Organization. The firm has over 160 employees, of whom about 11 reside abroad. Approximately 100 employees work in production and the remaining work in administration and service (including management).

Objectives for Quality Management

The objectives of quality management are described by Milliken's Business Philosophy and the related Business Policies. The most important of these policies is the quality policy.

Milliken's Quality Policy

- Milliken is dedicated to continuous improvement of all products and services through the total involvement of all associates.
- All associates are committed to the development and strengthening of partnerships with Milliken's external and internal suppliers.
- Milliken will continually strive to provide innovative, quality products and services to enhance its customers continued long-term profitable growth by understanding and exceeding their requirements and anticipating their future expectations.

HISTORY OF QUALITY MANAGEMENT

It Started with Measurements

"Before you start changing anything, find out where you are now." Or, put another way, "The quality process starts with measurements." (A measurement is derived from data, and the results or measure are key pieces of information for the organization.) What the Danish organization was being told, in fact, was that the firm's future operations should be based on fact, not beliefs and opinions. There are many types of measurement, and some of the most important are concerned with employee safety, timely delivery, customer satisfaction and failure costs, employee satisfaction, credit memos, employee education, wasted materials, composition of inventories, and the number of complaints about goods and services.

The Company Barometer

"The Company Barometer" is the annual customer satisfaction survey, which also compares Milliken's customers' satisfaction level with that of its biggest competitors. This is the most important statistic for the company.

Milliken has found that it is possible to achieve an improvement of almost 50% in the first year simply by measuring and following up regularly, in more or less every area.

The Suggestion System

Milliken Denmark has had a formal suggestion process, the OFI process (Opportunity For Improvements), since 1984. The system deals with more than 4,000 opportunities for improvement per year. The system has been found to be indispensable, as well as complementary to the Error Cause Removal (ECR) process. The OFI system has proved to be an excellent vehicle for transmitting employees' ideas to the ECR teams.

Zero-Defect Philosophy and ISO 9001

It was only upon achieving ISO certification in 1988 that Milliken finally found a "tool" to help in preventing defects. This tool is the "Correction Report," the purpose of which is to describe defects systematically and ensure that their causes are eliminated. More than 1,000 correction reports have been written since then, each of which has eliminated at least one source of error.

The European Quality Award

Milliken European Division, of which Milliken Denmark is a part, participated in the competition for the European Quality Award in 1992, when it won a European quality prize. In 1993, it won the award.

Summary

Milliken's quality process has been under way since 1984, and several learning points have been recognized. One is that the quality process required patience and endurance. Another is that progress is often of the "two steps forward, one step back" variety, and that the best remedy when one starts to lose heart is to compare the firm today with what it was a couple of years ago. Milliken has learned that there is no alternative to the quality process.

The Milliken Philosophy

"At Milliken our culture is firmly based on 'the Pursuit of Excellence.' As well as setting standards for the quality of our products and services, this philosophy also embraces the quality of our people. Each individual is called an Associate, and each plays a vital part in achieving the Milliken goal."

"Milliken has been working with all aspects of 'Total Quality Management' for the past 13 years, and is now recognized as a leader in this field. This dedication and hard work was recognized when, in 1989, Milliken received one of the highest awards for quality: The Malcolm Baldrige Award, which was presented to Mr. Roger Milliken by the then-President of the United States, George Bush. After having received the European Quality Prize in 1992 Milliken Europe won the European Quality Award in 1993."

IT STARTED WITH MEASUREMENTS

Milliken Denmark's great fortune was that its parent company had a couple of years' lead and was thus able to set an example. The group management showed the necessary leadership and, from the start, one point was emphasized over and over again: "Before you start to change anything, find out where you are now." Or, put another way: The quality process starts with measurements.

What the Danish organization was being told, in fact, was that the firm's future operations should be based on facts, not beliefs and opinions. This was echoed by Peter Hørsman, managing director, who declared that, from now on, guesswork was out, adding that one measurement was better than ten opinions.

Employees' Safety

As can be seen from the Safety Policy, Milliken gives top priority to employees' safety. The firm's basic principle is that all accidents can be prevented. Every safety problem is followed up until an acceptable solution has been found. The following anecdote illustrates this.

Throughout the 1980s, the American division president paid Milliken Denmark an annual 1-day visit. The top management team members had about 8 hours to tell him everything about the business, their plans, visions, investment hopes, etc.

A few days before one of these visits, one of the cleaners slipped on some ice in the parking lot and broke her ankle. This episode took up 3 of the 8 hours of the president's visit, because he wouldn't let management get away with passing it off as an unfortunate accident. Management had to draw up plans to make sure that similar accidents could be prevented in the future.

Every 4 weeks for the past 12 years, management has sent reports about safety conditions to the group management, regardless of whether there has been anything to report. Milliken keeps statistics on accidents that result in sick leave and on minor accidents where sick leave hasn't been necessary. Milliken Denmark has also been required to keep statistics on irresponsible behavior that could have led to an accident.

The Company Barometer

The "company barometer" is an annual customer satisfaction survey. A sample of customers are interviewed by a bureau on a wide range of issues connected with how they rate collaboration with Milliken Denmark and its competitors. If the survey shows that Milliken's competitors have come out on top, or if Milliken has achieved a lower score than in previous years, then corrective measures are taken. *This measurement is regarded as the most important measurement of all.*

Milliken's experience shows that it is possible to achieve an improvement of almost 50% the first year, simply by measuring and following up regularly in more or less every area.

Customer Satisfaction and Hoshin Planning

The following seven business fundamentals are evaluated in the Customer Satisfaction Survey:

1. Quality
2. Cost
3. Delivery
4. Innovation
5. Morale
6. Environment
7. Support

Each of these seven parameters is evaluated on a scale from 1 to 10 and compared with those of the most important competitors. These measurements enter the strategic planning process (Hoshin planning process), in which strategic teams evaluate the results in order to start the process of establishing new objectives, targets, strategies, measures, etc., in order to secure continuous improvements in customer satisfaction. For this purpose, the teams use specially designed forms, which have proved to be a valuable tool in ensuring that the plans for improvement will have concrete results.

On each of the above parameters, customers have evaluated Milliken's quality between 8 and 9, and compared it to its biggest competitors. Milliken's overall quality level is approximately 0.8 higher. This is the definite proof that Milliken's TQM journey has been a very valuable one.

Keeping Delivery Dates

The third area that Milliken Denmark has paid particular attention to from the start of its quality process is keeping delivery dates. Every morning at 9 A.M., all office staff are informed about how successful they have been in fulfilling the delivery deadlines promised to customers. If there have been any delays, a full account of why they have occurred is given.

Each week Milliken has to report to the European division manager, and each month to group management in the U.S. Excuses such as "a subcontractor was responsible for the delay" don't count. The goal is 100% delivery on time and the "score" is over 98%. In the U.S., the manager of any company (there are over 50 companies in the group in the U.S. alone) that comes under 97% has to present himself at the monthly board meeting and explain the measures he has taken to prevent it happening again.

When analyzing the measurements of on-time delivery, Milliken has for several years used a control chart, which has been a valuable tool in determining when the delivery process is out of control.

Other Measurements

Apart from the areas mentioned above, about which Milliken's group management is especially concerned, Milliken Denmark measures many factors, including employee satisfaction, number of credit notes, employee education, the amount of time it takes to process designs, number of guests, wasted materials, composition of inventories, and the number of complaints about goods and services.

THE SUGGESTION SYSTEM (A KEY FORM OF INFORMATION)

Rediscovering the Suggestion Box

In 1987 Milliken rediscovered the much-maligned suggestion box. The company called the suggestion system its Opportunity For Improvement (OFI) process. All employees were invited to suggest ways of improving the firm. There were only a few suggestions to begin with. People didn't feel they had any contribution to make—not if they had to write it down and send it to management, at any rate.

People had all sorts of excuses, which really boiled down to their being unaccustomed to expressing themselves in writing: feelings of inferiority, and fear of making spelling errors or writing illegibly. The breakthrough came at a staff meeting, where the managing director convinced them that he really needed their suggestions. In addition, their suggestions would be a good way of letting him know what employees thought could improve the firm. And, since he was often away on business trips, and therefore couldn't see to all employees as often as he would like, he offered to share his secretary with them.

So twice a week, the management secretary made the rounds among all production employees, and the result was overwhelming. She got one good suggestion after the other on everything from organization of the workplace, through product improvement, to changes in machinery and processes, plus a lot of minor suggestions that, though not terribly significant, were undoubtedly good solutions to daily irritants.

After about a year, these biweekly visits became unnecessary—by that time employees had overcome their reluctance to put pen to paper. Today, the number of suggestions has grown to over 4,700 a year, and a whole organization has grown up to respond to them.

All approved suggestions are rewarded with a small prize, the size of which has only a minor relation to the importance of the suggestion for the company. Basically, Milliken believes that the OFI system is in the interest of all employees, ensuring a strong firm, giving them greater job security and the chance to earn more. This belief also reflects their conviction of employees' collective responsibility for the continuing existence of the firm. Milliken realizes that not all employees share this belief, but it is also realized that it is management's responsibility to help them appreciate this.

The Suggestion Process and Error Cause Removal

Milliken has found the OFI system indispensable in complementing the Error Cause Removal process. Many opportunities for removing root causes of error or for making process improvements are brought to the management's attention through the suggestion system. Forms are readily available for everyone throughout the plant and offices and many ideas that come up during discussions and process review activities are captured immediately on an OFI form.

The person submitting the OFI form (or the OFI committee) is asked to classify the suggestion into one of eight alternative OFI types equal to six of the seven business fundamentals evaluated in the Customer Satisfaction Survey (see section B1), and later on used as input to the Strategic

Planning Process, called Hoshin Planning, plus Safety and Energy. The eight OFI types are shown below:

1. Quality Improvement
2. Cost Reduction
3. Delivery
4. Innovation
5. Safety
6. Morale
7. Environment
8. Energy

The OFI types' close similarity to the seven business fundamentals complements the Hoshin planning process. It happens frequently that the management asks the whole organization to support a certain activity by contributing ideas. An example of this is Milliken's periodic drives to improve safety in the plant and to reduce the environmental impact of its activities. The organization responds with many ideas on how this could be accomplished and these ideas in turn feed into the Error Cause Removal team activity.

The OFI process plays a very important role as a communication tool on a smaller scale, as well. When Milliken has a problem area, management establishes an Error Cause Removal (ECR) team of four to six employees to address the problem. Many employees outside the team are normally affected by the problem, or they are aware of it, and they often have opinions on what the cause is and how it can be solved, regardless of whether they are members of the team. The OFI system has proved to be an excellent vehicle for transmitting employees' ideas to the ECR teams.

The Suggestions Received

The suggestions received can be classified into four categories:

1. About 20% are rejected as unsuitable.
2. About 30% are minor suggestions.
3. About 45% are good suggestions that can make a difference to day-to-day activities.
4. About 5% are really good suggestions that can make all the difference to competitiveness.

The annual savings cover only the "visible quality costs." The gains from these ideas have, among other things, financed construction of a new factory and an administration block with a modern auditorium.

The suggestion system runs very smoothly. For several reasons, the program has become a natural part of daily work. One reason is that new employees are given a sponsor. The sponsor is usually a manager from another department whose responsibility is to make it easier for the new associate the first couple of weeks in the new company and to help him or her to submit his or her first suggestion (OFI). The sponsor simply asks if there is anything the new associate has observed that he or she thinks may present an opportunity for improvement.

Milliken's belief in the importance of helping the new employees become acquainted with the suggestion system can be understood by reading the following quotation from Mr. Ejvind Jensen:

> A company should be especially attentive to suggestions from new employees, because a new employee's power of observation isn't yet reduced by routines and principles. It is very important that the commitment · from new associates isn't suffocated with remarks like "we don't work this way." Instead we point out that we expect to hear from them what we can do better according to their point of view. This goes for all associates, whatever their function or position in the company.

This attitude and the resulting "sponsorship for suggestions" is an example of Milliken's "Leadership for Suggestions."

Another example showing Milliken's "Leadership for Suggestions" is that managers (including top management) are also very active in submitting suggestions for improvement to the OFI system, although they are not rewarded for doing so. Only ordinary associates are rewarded for participating in the OFI program.

Committee Production meets every afternoon at 1:00 P.M. The meeting last about 1 hour and deals with 10 to 30 OFIs. In the afternoon meeting, three associates from production participate voluntarily. Every 6–8 months, these three positions are vacated, and those who are interested are encouraged to apply as new members. It has become very popular to be a member of the OFI committee.

The administration of the OFI program is handled by the managing director's secretary, who is responsible for the registration of the OFIs and has a specially designed computer database to assist her.

Every month a list is produced, containing all the associates' scores (number of OFIs and number of points). The list is distributed to all departments and posted on the notice boards. A project manager and a colleague are more or less permanently engaged in implementing the suggestions to the mutual benefit of both employees and the company.

The success of the OFI system is the result of "Leadership for Suggestions." One example is the quick response to suggestions. The OFI committee holds meetings every day in order to be able to give quick feedback. The goal is to give feedback within 24 hours, but if an OFI is circulated for comments, the person who submitted the suggestion receives a provisional response within 72 hours.

EXPERIENCES WITH ISO 9001 (KEY INFORMATION)

ISO Certification Failures

The last months of 1988 were a milestone in the history of Milliken Denmark. This was when the quality assurance system finally obtained ISO 9001 certification. Twenty office staff members had spent 4,000 working hours, over 10 months, documenting the system.

The documentation process was extremely important, since it gave Milliken the opportunity to examine every nook and cranny of the company. There were overlapping areas of responsibility in several parts of the firm, and, perhaps worse, other areas where there was no precisely defined responsibility at all. Coordination at management level during this phase undoubtedly speeded the quality process along.

The management group thought that it would end up with "quality management," but it found that ISO certification is more about the quality *of* management. It understood that it had to demand integrity and responsibility from all managers if the certificate was to be more than just a pretty diploma on the wall.

Management underestimated the interest and involvement of shop floor workers in the documentation phase. Management was so focused on its goal that there wasn't time to tell employees what sort of certificate Milliken was trying to win, what lay behind such a certificate, or how they would be affected by it. Employees were fobbed off with assurances that, as far as they were concerned, it would be "business as usual," the only difference being that now it would be in the name of a formal quality assurance system.

Once the press conferences and receptions were over, management was accused by employees of having pulled a fast one. Management's story was equated with the tale of the Emperor's New Clothes. This seemed unfair to the management group, of course, because during the lifetime of the company, Milliken had won over 40% of the European market for rental mats.

THE EUROPEAN QUALITY AWARD

Milliken European Division, of which Milliken Denmark is a part, participated in the competition for the European Quality Award in 1992, when it won a European quality prize. In 1993, it won the award.

The self-assessment framework of the EQA model helped Milliken a great deal in obtaining a much better overview of all its quality-related activities, including the management process.

One of the key lessons the company learned from the 1992 competition was that it needed to reassess its process management. Members of management learned the importance of following the Plan-Do-Check-Act cycle for continuous improvement and to ask themselves whether a process itself was designed to generate the results they were striving for.

COMPANY

Pal's Sudden Service

PRODUCT/SERVICE:

Food and beverage items provided through a quick-service, drive-through restaurant chain

LOCATION:

Kingsport, Tennessee

NUMBER OF EMPLOYEES:

325

KEY PERSONNEL:

- Pal Barger, CEO
- Thom Crosby, President and COO
- 15 Store Operators/Owners
- Case Study Contributor and Quality Consultant: J. Michael Lewis

KEY USES OF INFORMATION:

1. Strategic Planning
2. Customer Focus and Market Research

3. Benchmarking
4. Empowering Employees for High Performance
5. New Product, Service or Process Introduction
6. On-line Quality Control
7. Continual Improvement
8. Management Review of Progress and Performance

ORGANIZING FRAMEWORK TO CREATE KNOWLEDGE:

Management Information System and the Malcolm Baldrige National Quality Award criteria

AWARDS AND SIGNIFICANT RESULTS:

1. 1995 Tennessee Quality Award for Commitment
2. 1996 Tennessee Quality Award for Achievement
3. 1998 Malcolm Baldrige National Quality Award Applicant
4. Fastest service times, highest health scores, lowest staff turnover, highest % profit, and highest productivity in the market area

INTRODUCTION AND BUSINESS OVERVIEW

Pal's is a small, privately owned, quick-service restaurant (QSR) company competing directly with major internationally recognized chains to offer food and beverages ready for customer consumption. Unlike other businesses, a restaurant consists of manufacturing, service, and retail components all combined into a single business operation. The manufacturing plant feeds raw materials into scheduled processing, assembly, and packaging steps to produce food items ready for consumption. The service component receives and fills customized orders in a responsive, friendly, timely, and accurate manner. The retail component provides a carefully designed, displayed, and priced menu of items for direct sale to buyers. Arriving unannounced and at random times, the customers expect immediate availability of food products for which they make customized requests. To achieve success, this multi-faceted business requires the effective use of a diverse management system with capable processes and tools common to a variety of companies.

Pal's, which operates 15 quick-service restaurants, was founded by Fred "Pal" Barger in 1956 with an overall target market consisting primarily of eight small to mid-sized cities in the geographical area of Northeast Tennessee and Southwest Virginia. The product line consists of a focused group of food and beverage items with a unique flavor profile designed to meet customers' taste requirements. The core menu consists of hamburgers, hot dogs, sliced ham and chicken sandwiches, french fries, breakfast biscuits with country ham, sausage and gravy, coffee, orange juice, milk shakes, iced tea, and soft drinks.

By listening to customer needs and benchmarking successful restaurants nationwide, the company designed and employs a drive-thru store concept. Pal's stores are designed to provide a facility with low start-up cost, an ultra-efficient operating process capable of selling the highest quality hamburgers and hot dogs at the most competitive price, and the quickest (i.e., "sudden") service times in the market. In addition to producing both lower cost and faster service than any of Pal's direct competition, the new stores featured a better trained staff dressed in an upscale uniform and a management system based on the principles of Total Quality Management. The overall effect has pleased the customers so well that Pal's has experienced a steady 12-year period of expansion and sales to meet customer demands and generates over $10 million in sales each year.

In the quick-service restaurant business, the facilities and equipment must be capable, reliable, and sanitary; zero variability, zero shutdowns, zero delays, and zero health problems are the requirements for operations.

Cleanliness, sanitation, safety, waste disposal and labor practices at Pal's are regulated by a variety of local, state, and federal agencies.

Pal's has performed extensive market research to pinpoint customer requirements: convenience, ease of ingress and egress, easy-to-read menu, simple accurate order system, fast service, and wholesome food. Key customer requirements have been translated into Key Business Drivers: quality of products and service, service, cleanliness, value, people, variety, and speed. Each requirement is linked to clearly defined operational processes, procedures, and systems that are continually monitored to ensure that Pal's meets customer requirements. Pal's maintains ongoing communications links to its customers so that it can listen to how well customers think Pal's is meeting their needs and to learn if customer needs are changing or if new needs have arisen.

Pal's has four key suppliers who provide 97% of all raw material vendor purchases. Pal's views its vendors as a key stakeholder group and gives them a priority equal to that of customers by maintaining an ongoing, open communication link. In addition to the cost, reliability, safety, and consistency of supplies and raw materials being of paramount importance, Pal's cannot meet its operational and customer requirements unless its suppliers operate with no surprises, no delays, and no variability.

Pal's competes primarily with franchise locations of major international and national fast food chains. Within its service area, Pal's with 15 stores has a 15.4% share of the QSR sandwich market. Pal's major competitors and their respective market shares are McDonald's (20.8% total market share); Burger King (15.9%); Taco Bell (14.1%); Hardee's (13.8%); Wendy's (12.7%); and all others (7.3%). Pal's competes in a small market head-to-head with these much larger companies that enjoy the huge competitive advantage held by national chains of having a louder marketing voice. To stay in business Pal's must focus on delivering excellence hour-to-hour, day-to-day, year-to-year, taking TQM seriously. This business is highly regulated and highly visible to the public with no place to hide problems, no way to cover up mediocrity, and no room for error. The negative publicity afforded health problems and the power of word-of-mouth advertising in a small market are unforgiving.

LOGIC

Pal's employs a unique business strategy among QSR companies that has guaranteed its ongoing financial success. Strategic components that differentiate Pal's from its competitors include

- Total reliance on a **TQM Customer-Focused Management Process**
- Extensive use of benchmarking to find and adopt **Best Business Practices**
- **Slow, Calculated Business Growth** based on tested and proven operational capabilities, customer acceptance, and company-wide standardization
- Expansion with the drive-thru facility designed for operations that produce **Sudden Service**
- Thorough **Training, Empowerment and Reduced Turnover** of a traditionally transient workforce
- Capitalizing on **Health and Sanitation Regulations** to create a distinct competitive advantage
- **COO and Operator Ownership** in the business with compensation directly tied to financial results and
- **Business and Process Leadership** centered around carefully screened and capable Store Operators who are empowered with responsibility, accountability, and authority both as active members on our Leadership Team and as an ever-present mentor and coach on the store's Process Team.

This business strategy is designed into Pal's Business Excellence Process, which is fueled by its Management Information System. Pal's assuming a leadership role in the restaurant industry by modeling business success produced through a stakeholder and information-based management process. Pal's makes extensive use of listening techniques, self-assessment, enlightened benchmarking, and on-going piloting of new product, service, and process ideas to provide the information needed to produce business excellence. Pal's is setting the example of responsible management by continually improving its sudden service drive-thru store concept, streamlined operational processes, and strategies for delighting customers. This system, although common in the manufacturing sector and becoming more common in the service sector, is a unique approach to business in the food industry which has traditionally been driven by aggressive top-down cost control and comprehensive promotional strategies.

Pal's Leadership Team (CEO, COO, and Store Operators) has adopted a **Business Excellence Process** to serve as the overall leadership system for guiding and aligning business decisions and actions throughout the organization. Leadership activities in support of the Business Excellence Process provide the framework by which company senior executives initiate and deploy plans to properly address the desired values, company directions, performance expectations, a focus on customers and other stakeholders, learning, and innovation. Included are six main interactive components: the Strategic Planning Process, the Benchmarking Process, the New Product/Service/Process Introduction Process, the On-line Quality Control Process, the Continual Improvement Process, and the Communication/Feedback Model.

The **Strategic Planning Process** is used to set direction and pursue future opportunities for the business while taking into account the needs and expectations of all key stakeholders. By defining, aligning, reviewing, and maintaining a Corporate Mission Statement, Vision Statement, Action Plan, Key Business Drivers, and Code of Ethics, Pal's maintains clear values, high performance expectations, and a keen focus on stakeholder needs. These strategic planning outputs are communicated and interpreted at each organizational level to define both organizational and individual responsibilities and identify opportunities for learning and innovation.

The **New Product/Service/Process Introduction Process** is Pal's systematic approach for developing new or modified products, services, or processes. It includes clear customer input, thorough market research, detailed design, extensive testing and analysis, heavy involvement and feedback by employees, and calculated rollout to insure internal capability and customer acceptance.

The **Benchmarking Process** complements the Continual Improvement Process by guiding Pal's efforts to make meaningful competitive comparisons, to identify and adopt best practices needed to achieve high performance expectations, and to achieve defined directions for the future.

This emphasis serves as a constant reminder that the current level of performance can always be improved and provides many of the ideas for reaching stretch goals. It also helps pinpoint the primary areas for organizational and individual opportunities for learning and innovation.

The **On-line Quality Control Process** is used to apply best practices on a consistent basis to achieve high performance and excellence in operational and support processes. General Staff training and coaching are applied to develop capabilities to meet strict operational standards and performance targets. In-process and end-product measures are monitored on a real-time basis to identify process upsets that require immediate attention.

The **Continual Improvement Process**, featuring a Plan-Do-Study-Act cycle (PDSA), is used to define standards of excellence, to maintain a focus on customers and other stakeholders, and to identify, plan, and execute improvement projects. The Malcolm Baldrige Criteria is used in conjunction with the Continual Improvement Process as a self-assessment tool driving Pal's toward overall excellence, accelerating implementation of the business vision, and maintaining an emphasis on Pal's processes for delighting the customer.

The **Communication/Feedback Process** provides a means for communicating and deploying directions and plans to all organizational levels and all stakeholder groups. It was first utilized to establish the original versions of Pal's Mission, Vision, Action Plan, and Code of Ethics, which were drafted by the Leadership Team. Through communication and feedback, the Leadership Team is able to more confidently set direction by giving special attention to the principles upon which the company was founded, the needs and expectations of all key stakeholders, and the operational and business challenges facing the company. The feedback loop is employed to allow the entire organization and all stakeholders to participate in the Pal's leadership system. Each leadership component at Pal's is communicated, discussed, interpreted, improved, and re-communicated at each level of the organization (both horizontally and vertically) and with each stakeholder group (not just employees and customers).

The Pal's leadership system (i.e., Business Excellence Process) relies heavily on feedback throughout all company organizational levels and among all stakeholders to ensure that the leadership components will add value to everyone who has a stake in Pal's business success. In addition, it produces directions and plans which are closely aligned with everyone's needs, roles, and capabilities. The "DO" and "STUDY" steps of our Plan-Do-Study-Act cycle (Continual Improvement Process) are significantly driven by organizational and stakeholder feedback, greatly enhancing the quality of information flowing throughout our organization and our stakeholder network. This healthy flow of information has greatly improved Pal's ability to deploy direction, plans, and actions and to achieve desired results. Extensive use of stakeholder feedback has significantly improved the commitment to and validity of Company direction and plans, the alignment of all stakeholders to the company directions, and the magnitude and rate of performance improvement realized for all of Pal's Key Business Drivers.

Ongoing communication and feedback between all levels at Pal's provide a comprehensive view of the strategic needs and forge commitment to near- and long-term goals and strategies. Peer reviews, review of management performance, and annual staff surveys are used to measure how well these responsibilities are being executed and how well the culture is working. These measures are also used to assess the effectiveness of the leadership system. The goal is for everyone to rapidly achieve understanding of, develop a commitment to, and get meaningfully involved in the vision, direction, and improvement plans which are guiding Pal's toward excellence in all areas of performance. Early and frequent involvement in the design of goals, plans, processes, and procedures provides constant information flow, keeping everyone informed and linked to the overall direction. As the organization identifies further ideas for improvement opportunities, they are reviewed and assessed by the Leadership Team.

Although the Leadership Team assumes the responsibility for initiating the Business Excellence Process, ideas for setting new directions and pursuing future opportunities frequently originate from employees (and sometimes stakeholders) outside the Leadership Team. Each senior executive spends a portion of every day actively seeking input from employees, current customers, potential customers, suppliers, and other stakeholders to gather inputs used in the Business Excellence Process and data used to assess the effectiveness of the leadership system. This shared leadership approach challenges all of Pal's employees to continually learn and improve their skills, capabilities and contributions. In addition, process ownership and continual learning are driven throughout the organization by ongoing training in Best Business Practices. Managers coach and mentor the staff to reinforce a culture of responsible self-control and empowered decision-making.

COLLECTION, INTEGRATION AND ANALYSIS OF DATA

Pal's Leadership Team has carefully designed a system for the collection and analysis of information and data that feeds and interacts with each component of its Business Excellence Process. The Pal's Management Information System (MIS) is also used to guide the selection, management, and effective use of information and data to support key operational processes, action plans, and performance management system.

The information and data collected at Pal's are used to produce a Balanced Scorecard of Core Performance Measures (Table 1) and to support the management and continual improvement of overall company and operational excellence. The balanced scorecard links in-house performance measures and data charts for each Key Business Driver with the influencing key operational processes and strategic action plans. In addition, measures are provided to assess the impact of performance for corresponding business support processes and supplier processes.

After data selection by the Leadership Team, the Communication/Feedback Process is employed to verify and further specify information and data needs throughout the organization. After the communication, review, feedback, review, and modification process for data selection is complete, a final version of the balanced scorecard for organizational dissemination and use is designed.

Information Frequency Tables determine the specific store measures, the frequency of reporting, and the level and frequency of review. The Pal's MIS collects data directly from each store and returns performance reports with both store and overall company data analyses. These two sets of performance data ensure a balanced scorecard is received at each level of the organization to provide specific, timely information that each Pal's employee (from hourly staff to senior executives) needs to monitor, evaluate, control, and improve performance for each process.

The ability to provide meaningful, timely, reliable, and user-friendly information and data is assessed by key users throughout the data selection, collection, analysis, reporting, and review process. Pal's begins considering user requirements during the data and information selection stage with pre-defined user criteria. The data selection, collection, and reporting criteria are based upon

- meeting stakeholder requirements
- linking to the Key Business Drivers
- providing a balanced scorecard
- creating reliability
- providing rapid access
- allowing rapid update
- supporting continual learning, and
- supporting continual improvement

Pal's also ensures that user requirements are more closely met by gathering and processing most input data for the MIS at its source (in the

Table 1   Pal's Balanced Scorecard of Core Performance Measures

KBD	Measure	DT	FQ	AC	Goal/Target
Quality	Complaints	C	D	1	< 60 per Year per Store
	HACCP Temperature	O	D	1	100% On-Target
	Customer Counts	C	D,M	1	> 8% Increase
	Market Share	C	A	2,3	17%
	Process Measures	O	D	1	All processes within control limits
Service	Service Speed	O	SA	1	25 seconds or less
	Complaints	C	D	1	< 24 per Year per Store
	Accurate, Fast	C	A	2,3	Customers rate us Best of Class
Clean	In-House Health Inspection Scores	O	M	1	98% Score
	State Health Inspection Scores	O	SA	1	98% Score
	Pre-Operational and Operational Checklists	O	D	1	No Deviations from Standards
Value	Average Sales Check	F	D,M	1	> $3.50
	Average Hourly Wage	F	M	1	> All Competitors
	Sales Per Labor Hour	O	H	1	> $30
	Sales	F	M,A	2	> $13.0 Million Sales
	Profit	F	M	2	> 15%
People	Staff Surveys	O	A	1	Higher Score than Prior Survey
	Accidents	O	SA	1	< 15 Per Category
	% Staff Turnover	O	M	1	< 120%
Speed	Service Delivery Time	O	SA	1	< 25 Seconds

LEGEND: KBD=Key Business Driver. **DT**=Data Type (C=Customer, O=Operational, F=Financial) **FQ**=Frequency (H=Hourly, D=Daily, W=Weekly, M=Monthly, Q=Quarterly, S=Semi-Annually, A=Annually) **AC**=Access (1=All Team Members, 2=Leadership Team Members, 3=Third Parties)

store). The MIS generates store-level and company-wide reports on sales, customer count, product mix, ideal food and material cost, and turnover rates, which are provided to the store level where the data is reviewed and its accuracy validated. Feedback collected from the stores by the MIS is used to develop data reliability reports which are transmitted to Operators via fax to confirm the validity of the monthly performance reports. Furthermore, standardized software and data collection/entry procedures at each store allow the information to be entered into the system one time to help enhance its reliability.

In addition to these approaches to achieve data reliability, continual review and evaluation are conducted by users who provide feedback to the Leadership Team regarding the quality of information and data provided through the MIS. This rigorous and complete feedback system ensures that user requirements are addressed and needs for improvement are identified. The Continual Improvement and Benchmarking Processes are then applied to identify data improvement ideas and to initiate pilot efforts.

Principle financial and non-financial performance measures are integrated and analyzed for Pal's Key Business Drivers by using the Data Integration and Analysis Format shown in Table 2. Performance and capability data from all parts of the company are easily integrated because each of Pal's 15 stores has a common core set of performance measures including customer-related, operational, human resource, competitive, and financial performance. Data is provided for analysis and review from standardized data collection procedures and with standardized operational definitions/equations. In fact, each store provides standard data for a company-wide balanced scorecard to the centralized MIS where overall company integration and assessment occur.

STRATEGIC PLANNING

Pal's relies on long-range visions and strategic planning to produce customer-driven quality, market-focused growth, operational excellence, and responsible corporate citizenship and to guide ongoing decision-making, resource allocation, and company-wide management. The Pal's Leadership Team members are the stewards of the process for strategic planning and are responsible for the deployment of its strategies, goals and plans.

The Pal's Strategic Planning Process provides a disciplined and structured approach for setting strategic directions to strengthen business performance and competitive position. While using this process, the Leadership Team considers a variety of data from all levels of the organization and from primary stakeholders to establish strategies and plans for long-term customer satisfaction, business excellence, expansion, growth, and profitability. All strategic data analyses and interpretations are carefully evaluated against Pal's own operational requirements, capabilities, and available capital before selecting strategic plans. In addition, new strategic directions are assessed against the company's Mission, Vision, Action Plan, Code of Ethics, and existing directions. They are also tested

Table 2   Data Integration and Analysis Format

	Measures	Unit of Measure	Type of Analysis	Mgmt. Review/ Control Actions
Customer- Related Performance	Complaints	Number by Type	Cause & Effect	Identify root cause.
	Customers	Number of Transactions	Trend Comparative Cause & Effect	Identify trend (+/–). Verify validity. Identify root cause.
	Market Share	% of Market	Comparative	Assess satisfaction.
	Fast, Accurate Service	% Responses	Comparative	Define customer needs. Assess satisfaction.
	Supplier Satisfaction	Order Size, Payment Date	Comparative	Assess performance vs. goals/targets.
Operational- Related Performance	HACCP Temperatures	Temperature (° F)	Trend	Identify processes in non-compliance.
	Process Measures	$ Dollars	Cause & Effect	Identify root cause.
	Service Times	Seconds	Comparative	Assess performance vs. goals/targets.
	Customers	Number of Transactions	Comparative	Assess performance vs. hourly goals.
	In-House Health Inspections	Scale (1–100 Scale)	Trend Cause & Effect	Identify trend (+/–). Assess violations. Identify root cause.
	State Health Inspections	Scale (1–100 Scale)	Trend Cause & Effect	Identify trend (+/–). Assess violations. Identify root cause.
	Staff Surveys	Scale (1–5 Scale)	Trend Cause & Effect	Identify trend (+/–). By subject. Identify root cause.
	Turnover	%	Comparative	Assess performance vs. previous year.
Competitive Related Performance	Market Share	%	Trend	Identify trend (+/–). Identify root cause.
	Sales	$ Dollars	Trend Comparative Cause & Effect	Identify trend (+/–). Verify validity. Identify root cause.

Table 2   Data Integration and Analysis Format (Continued)

	Measures	Unit of Measure	Type of Analysis	Mgmt. Review/ Control Actions
	Turnover	%	Comparative	Assess performance vs. competitors.
	Profit	%	Comparative	Assess performance vs. competitors.
	Service Speed	Seconds	Comparative	Assess performance vs. competitors.
	Health Scores	Score (1–100 Scale)	Comparative	Assess performance vs. competitors.
Financial Related Performance	Sales	$ Dollars	Trend Comparative Cause & Effect	Identify trend (+/–). Verify validity. Identify root cause.
	Profit	%	Comparative	Assess performance vs. goals/targets.
Market Related Performance	Market Share	%	Trend Comparative	Identify trend (+/–). Assess performance vs. competitors.

for credibility, reasonableness, and commitment through Pal's Communication/Feedback Process with employees and stakeholders before finalizing any change in or addition to direction.

Customer and market data from a variety of sources are taken into account by using them to interpret customer needs and requirements, to project market trends, to establish new strategies that will delight customers and sustain Pal's competitive advantages, and to check the impact of previous strategies and plans on customer satisfaction. These customer and market data sources include Market Research Studies, Marketing By Wandering Around surveys, and Customer Complaint data.

Competitor information from Pal's benchmarking process is used extensively to identify market and industry trends, industry and competitor capabilities and Best Business Practices, competitor strategies, potential competitor reactions to our strategies, and promotional and technological improvement opportunities. These projections for the competitive environment are essential to detect and reduce competitive threats, to shorten reaction times for change, to identify key improvement opportunities, and to minimize the need for major strategic redirection which can easily stem from a lack of awareness of external factors and trends.

In addition, a key competitive factor in the quick-service restaurant business is pricing, which must be carefully aligned with market trends to maintain sales volume. Strategies for cost control are a critical component of the overall strategic plan to achieve desired financial results at anticipated price levels, rather than simply planning to set prices to cover costs. Pal's Financial Planning Process is a key element of this part of the Strategic Planning Process.

Strategic risk data are analyzed in several areas of concern using the following questions to assess potential value added or threats for all stakeholders:

Market Risk
Where are new competitors locating?
What are the competitor's pricing, products, and service strategies?
What new market niches are emerging?

Financial Risk
What is the return on investment?
How do capital requirements compare to both short- and long-term potential gains?

Technological Risk
Are any of our technologies becoming obsolete?
Have any new hazards or risks been discovered with our existing technologies?
Are there technologies from other industries that could give us a competitive edge?
Are they cost effective?
Are our competitors developing or using any new technology?
Will the technology become an industry standard?
What training will be needed by our staff?
What safety risks are associated with it?

Societal Risk
What are the current health implications and issues?
Will this plan affect the local community (i.e., traffic, zoning, future development plans)?

The issues that emerge from this analysis are prioritized and compared to existing long-term goals. Depending on the value added, either immediate changes are incorporated in the current strategic direction with plans using Pal's pilot implementation model, or a plan temporarily monitoring and analyzing the target situation on an ongoing basis is established.

Financial risk is given special consideration with a comprehensive Financial Planning Process including annual assessment, budgeting, forecasting, and decision-making to guide short-term cash flow and long-term profits. Detailed cost projections are made for each store to accurately manage profit and loss, cash flow, expansion and capital investment, and daily financial decision-making. Actual financial results are closely monitored versus plans to determine when and how to make appropriate adjustments to strategic plans and improvement actions to achieve maximum long-term financial results at minimum financial risk.

Human resource planning is an integral part of Pal's strategic planning process. Once business and operational strategies and Key Business Drivers have been identified by the Strategic Plan, Human Resource (HR) strategies and plans are developed by the Leadership Team to train and educate personnel to build staff capacity for success, to improve work/job designs and work areas, and to build a better culture for excellence and employee well-being. These HR plans are developed based on input data and analysis from

- Assessment of current performance levels
- Interpretation of HR needs for achieving short- and long-range strategic goals, plans, and directions
- Evaluation of the annual staff satisfaction survey which includes questions on staff motivation, training and career development, work assignments, communication, and involvement
- Use of the Continual Improvement and Communication/Feedback Processes on a variety of measured performance data for the People (a Key Business Driver) during monthly Management Review Sessions, and
- Evaluation of financial and operational performance data analyzed at the store and company levels looking at the relationship with human resource plans and results

Anticipated needs and capabilities for human resources, technology, research and development, innovation, and business processes are identified by analyzing internal capability data against industry trends, technology developments, and performance advances. This approach allows Pal's to look for ways to develop synergistic combinations that can be used to help define dominant positions to occupy within its given market. Using the existing drive-thru concept, sudden service criteria, and process efficiency/excellence capabilities as a reference for comparison, ideas for planned experiments, performance assessments, and new learning tools are identified. Plans to assess these ideas on a small scale through Pal's New Product/Service/Process Introduction and Continual Improvement Processes are developed and included in the strategic direction to ensure human resources, technologies, and processes are continually improved and fully capable of meeting new and emerging customer requirements in the future.

Pal's four major suppliers, which provide 97% of its raw materials and supplies, are interviewed semi-annually to analyze how they perceive their own competitive market positions, their own process and performance capabilities, the quality and availability of their supply lines, their projected product cost trends, the impact of any new technologies in their industry, and any risks that may threaten their business and their ability to deliver quality and value in a timely manner to Pal's. Using the Strategic Planning Process, the Leadership Team projects Pal's future needs against current and anticipated supplier capabilities and their capacity to develop plans which support Pal's strategic direction. In addition, Pal's solicits supplier information about the capabilities and strategic plans of Pal's competitors and identifies any opportunities to include its suppliers in improvement initiatives.

The action plans and related human resource plans that Pal's has derived from the company's overall strategy are organized around Key Business Drivers identifying performance requirements, process measures, annual and long-term goals, and action plans to achieve company strategies. Key performance requirements and measures are determined by the Leadership Team at its monthly meetings. The operational goals are included in each store's business plan and operating budget. Goals are communicated to the entire staff by workplace scoreboards, meetings, training sessions, memorandums, and one-on-one discussions detailing how their performance in running their process will impact the planned strategic goals. In addition, meetings are held with Pal's four key suppliers to confirm their understanding of Pal's strategic plans and how their capabilities and improvement plans are linked.

Once strategic direction is defined and deployed, project teams are assigned and project plans are communicated again using the Communication/Feedback Process. For any action plans that require process or product modifications or introductions, the project team and involved staff are trained and plans are pilot-tested for implementation through the New Product/Service/Process Introduction Process.

CUSTOMER FOCUS AND MARKET RESEARCH

The priority of customer and market focus at Pal's is reflected in menu design (food items offered), flavor profiles (taste), pricing strategy, sudden service objectives, customer delightment vision, packaging designs, customer contact standards, store design, complaint resolution policy, sanitation standards, and dress code. Pal's ability to understand the voices of its customers and the market ultimately determine its business success. Pal's has therefore placed a major emphasis throughout its history on establishing a world-class process for listening, learning, and understanding what its customers and market expect and for determining how satisfied they are with the products and service Pal's delivers to them.

Pal's market research seeks comparative information relating to key issues such as what customers like or dislike about Pal's or specific Pal's competitors, and reasons why a particular restaurant is selected as a favorite. The comparative data is analyzed for trends and shifts in buying patterns to identify needed changes in our business strategy and to target other potential customer groups and future markets. From these analyses, Pal's identifies and applies different strategies and approaches for listening and learning to its various market segments.

Pal's listening strategies allow it to maintain continuous contact with customers and obtain valuable information directly from them to identify current and future needs. Product and service features are determined by listening to these wants and needs, benchmarking/evaluating competitor products and services, evaluating supplier products and capabilities, and ensuring Pal's has the technology and capabilities to meet these needs. Analysis of sales data and customer input from all customer listening/learning sources help pinpoint the relative importance of these features.

Pal's Market Research is a comprehensive study of fast food opinions, habits, and perceptions. To ensure objectivity, we contract with an outside service, Creative Energy Group, to conduct the survey, analyze the information, cross tabulate the results, and present the information by market segment to indicate customer needs.

Marketing By Wandering Around (MkBWA) puts Operators in direct contact with current and potential customers. During a 2-week period each quarter, the store Operators go door-to-door within a 3-mile radius of their stores, seeking face-to-face feedback on customer requirements and satisfaction levels. During MkBWA, operators listen one-on-one to customers asking questions designed to indicate buying trends and validate published trade data. The following questions are included:

- Do you eat at Pal's?
- Which products do you purchase?
- What is your opinion of Pal's service?
- What would you like to see different at Pal's?
- Which is your favorite quick service restaurant? Why?

Following each interview, the Operator records the answers to targeted questions. In addition, the Operator may leave the customer a mail-back form that can be returned later. This data is compiled company-wide, then returned to each store for review and analysis. The results are analyzed and discussed at Leadership Team meetings. Customer Complaints are tracked at the store level on an Opportunity Log then aggregated at the company level to indicate trends. This critical incidents information is used to understand key service attributes from the point of view of customers and frontline employees. Complaint data is analyzed against sales data and customer satisfaction results from Market Research to determine if its processes are meeting customer requirements and to ensure that the customer requirements (product and service features) Pal's has designed into its business are valid and working to meet expectations. If the data analysis indicates a shift in customer requirements and expectations, a pilot project is initiated with the Continual Improvement Process to investigate a cost-effective means of meeting those requirements.

The Pal's frontline staff training program includes intense instruction on effective listening skills. These skills are not only critical for high performance on the food preparation line (required for order accuracy, customization, and speed), but also are beneficial for gathering valuable customer information about needs, expectations, and satisfaction (as post-transaction feedback). Some of this information is recorded on the Complaint/Opportunity Log, some is shared in team meetings and performance reviews, and some is fed back through the Communication/Feedback Process.

Pal's is also able to realistically predict future trends using industry and competitor data. Understanding consumer tastes and buying patterns is a key element of customer knowledge which is influenced by many external and cultural factors creating rapid changes in buying patterns and tastes. Pal's uses information from industry trade journals relating to customer purchasing trends and future predictions. Membership and participation in national and local trade associations by Leadership Team members provide information on shifts in buying patterns, product features, equipment technology, and production methods.

All information regarding customer expectations and requirements is fed into the New Product/Service/Process Introduction Process to design product/service features and develop effective operational methods and capabilities.

BENCHMARKING

Pal's uses a systematic Benchmarking Process to determine best-of-class in practices and performance and to set stretch goals to reach and exceed best-of-class performance levels. The benchmarking process is part of the Business Excellence Process and feeds comparative data and best business practices to the Management Information System.

During regular performance reviews of the Balanced Scorecard, the Leadership Team assesses measured performance data against Pal's minimum performance expectation criteria for potential benchmarking partners (Table 3). This set of minimum performance expectations has been determined from competitive comparisons, Pal's past best performance, and Pal's strategic performance goals and serves as a reference for identifying strategically important performance gaps. The Strategic Planning Process identifies performance gaps and rates them according to their potential for creating an overall gain in competitive position and stakeholder value. Those performance gaps with the highest potential gain are given the highest priority for benchmarking and more extensive competitive comparison.

Once performance gaps are targeted, benchmarking partners are selected through on-line and paper-based research using the following criteria:

- Winner of Malcolm Baldrige or state-level Quality Award
- Recognized in their industry as best-of-class as demonstrated through publications and industry-level awards
- Recognized by a significant group of customers as best-of-market as determined by internal surveys and listening tools
- Outside of industry best-of-class as demonstrated through news articles, awards, and other types of recognition

All sources of comparative information and data (from within and outside the QSR industry and the Pal's market area) must either be a direct competitor or already achieving the levels of performance outlined in Pal's Minimum Benchmark Performance Requirements table.

Benchmarking Process information on the best performing companies in the industry and market is obtained through staff members sharing experiences from previous employment, observations made of direct competitors, and findings from formal benchmarking studies. Through restaurant industry research data from trade associations, literary searches, and access to the Internet, Pal's looks both inside and outside its industry for best-of-class performance and best-known method for continual improvement. Ideas identified are brought into Leadership Team meetings to determine if they have enough merit to pursue, and if so, a C-F Team is assigned to apply Pal's Benchmarking Process to find improved methods.

Leadership Team members maintain a "Hit List" of new technologies, equipment, or processes. They take this list with them to annual National Restaurant Association shows to see the most advanced technology and new systems that are or will be available in the industry. When enough

Table 3   Pal's Minimum Benchmark Performance Requirements

Key Business Driver	Measure	Minimum Performance Criteria
Value	Average Unit Volume	> $1.0 Million Sales per Store
	Average Check	< $4.50
	Profit	> 15% Pre-tax
	Food Cost	< 28%
	Product Pricing	< $3.00
	ROI	> 18% (all industries)
Service	Speed	< 5 minutes
	Accuracy	Best-Known (all industries)
People	Appearance	Best-Known (fine dining to QSR)
	Turnover	< 100%
	Professionalism	Best-Known (all industries)
Quality	Product Quality	Local Customers rate as Top Three Plus Top Three in Market Share
	Variety	Top Three in Market Share
	Value	Top Three in Market Share plus Avg. Unit Volume > $1.0 M Sales
	HACCP	Leader, Best-Known
Variety	Message (Advertising)	Top Five in Market Share plus Best-Known at message delivery (All Industries)
	Frequency	Top Five in Market Share plus Best-Known at message delivery (All Industries)
	Media Used	Top Five in Market Share plus Best-Known at message delivery (All Industries)
	Segment Targeted	Top Five in Market Share

favorable information is obtained, the new technology idea is assessed and piloted with the New Product/Service/Process Introduction Process to determine how it can be implemented to Pal's advantage.

Using the approach outlined in the Continual Improvement and Communication/Feedback Processes, the Leadership Team shares the results of competitive comparisons and benchmarking studies with the entire organization. By its nature, competitive and benchmarking information arouses a lot of internal curiosity and generates a lot of interest and motivation to determine the essence of competitors' processes and performance. Each employee is challenged to evaluate internal processes and/or performance against the benchmark information and competitive comparison data by exploring through innovative thinking how the study results could be used to improve Pal's performance and competitive position. The ideas generated from this challenge are fed back to the Leadership Team who review and analyze the ideas.

Using information about competitors, best-of-class performers, and best-known methods, Pal's develops stretch goals during management reviews and strategic planning sessions to push us toward high performance results. After analyzing and learning as much as possible from its competitor's processes and performance, Pal's uses its PDSA pilot and continual improvement approach to incorporate any learnings into its own processes to produce equal or better results. While reviewing data and benchmarks with best-of-class and best-known performers, Pal's forces itself to think "out of the box" and find innovative ideas and completely new breakthrough approaches for its processes.

EMPOWERING EMPLOYEES FOR HIGH PERFORMANCE

Employees throughout Pal's organization share common principles, visions, skills, and knowledge to achieve business goals and become more fulfilled individuals while doing so. Because of the challenge to produce excellence in operations (quality, speed, flexibility, service, and cost), human resources must be characterized by positive energy, employee well-being and self-control. Moreover, personnel must possess an ability to make the right decision at the right time and be dedicated to continuous learning, development, and growth. Given the young age and transient nature of the workforce at Pal's, Pal's must enable its people very quickly to develop and utilize their full potential on work assignments that are closely aligned with the company objectives.

Given the nature of the work in a quick-service restaurant and the work environment itself where staff meets the customer face-to-face, Pal's employs a work system that includes effective job designs and a flexible work organization to encourage individual initiative and self-directed responsibility. Through customer listening approaches and the Strategic Planning Process, Pal's has clearly identified key competitive factors (and Key Business Drivers) that pinpoint speed, accuracy, quality, and service as important performance expectations for the organization. Therefore, Pal's has focused human resource strategies, plans, job/work designs, and Human Resource Processes on developing capabilities that support these key competitive factors.

Pal's people, working mostly hands-on in labor-intensive and operator-paced jobs, are a critical element in its Business Excellence Process. They participate in the Strategic Planning Process through the Communication/ Feedback Process, they serve on Cross-Functional Teams to execute the Benchmarking and Continual Improvement Processes, and they are assigned as permanent members of Pal's Process Teams to apply the On-line Quality Control Process. Pal's extremely flat team structure requires that each staff member at the operational process level serve on a process team and be prepared to produce to the highest performance standards as the volume of customer arrivals demands it. High performance is accomplished through self-directed work designs organized along the processing, order-taking, service, and packaging line by process teams working with speed and flexibility while adhering to performance standards.

Work and jobs are designed to enable employees to exercise discretion, judgement, initiative and decision-making while strictly following best business practices that Pal's has determined to meet expectations for excellence and customer delightment. Pal's utilizes a simplified job classification format with only five job classifications (Store Operator, Assistant Manager, Opener, Closer, and General Staff) to allow it to assign work and make in-process adjustments rapidly to meet operational requirements and customer needs. In addition, existing systems and strategic plans are focused on providing critical process information at the job site to support real-time, self-directed decision-making. Information systems are also designed with input from Pal's process teams to provide the information necessary to support job and work designs and to ensure high performance. Team members monitor their own performance and adherence to process standards while utilizing the On-line Quality Control Process on each production and support process to ensure quality standards are met. Team members are all empowered to correct customer or quality standard problem on the spot.

Pal's drive-thru facility, which was designed to meet customer requirements for convenience, speed, and accuracy, was also specifically designed with an internal work layout around work processes to accommodate Pal's flexible, self-directed work and job designs. The processing and support technologies included were thoroughly evaluated for their capability to meet process, work, and job flow requirements. Pal's employees participated extensively in the facility and work station designs through Communication/ Feedback Process and Cross-Functional Team assignments using our piloting approach to standardization and continual improvement.

Pal's Communication/Feedback Process along with its work designs and team structure provides the basic foundation for formal and informal communication, cooperation, and knowledge and skill sharing across all business units (stores), across all organizational levels and with all stakeholder groups. Each component of the Business Excellence Process utilizes extensive communication with all levels of the organization and with major stakeholders (including customers) to identify issues, concerns, strategies, priorities, needs, plans, ideas, problems, learnings, and process improvements that impact business and operational performance. Likewise, extensive feedback is solicited, obtained and evaluated concerning changed directions, policies and methods to ensure validity, commitment, acceptance, and standardization in all that we do.

The efficient facility design and process layout puts process teams in a small work space requiring constant interaction, cooperation, teamwork and communication. The process must stay focused on meeting customer requirements and supporting an environment of trust, encouragement, and mutual commitment. The ability to communicate effectively throughout the process with both team members and with customers is a key requirement for high performance and therefore is carefully designed into the work, job, technology, training, and information systems.

Pal's flat organizational and Process Team structure provides continuous opportunities for communication among all organizational levels and categories of staff. Senior executives often visit the stores, while Operators work side-by-side on the production line within Process Teams with assistant managers and the general staff. These frequent opportunities for communication facilitate training, coaching, standardization, and the transfer of information concerning our operational processes and procedures.

In addition, sharing of information and organizational learnings is also facilitated by Pal's Cross-Functional (C-F) Improvement Team assignments where process team members from each store participate on assessment, benchmarking, learning, and improvement projects. Ongoing C-F Teams for the seven Malcolm Baldrige categories and Continual Improvement include members from each store who work on company-wide improvement opportunities and implement common learnings and standardized methods.

Throughout the business day, process teams from shift-to-shift communicate critical process and performance information during shift changes to ensure ongoing high performance. In additional, regular communications are maintained store-to-store via telephones and fax machines with established patterns and formats for information flow. If any store experiences problems with suppliers, products, or procedures, it will immediately alert all other stores that a problem is present thereby reducing the possibility of adversely affecting customer delightment. Strategic plans to provide a more real-time interactive management information system across the entire organization will further improve communications and shared learnings to better support Pal's high performance expectations.

Store Operators maintain a weekly Communication Log which lists important information about sales, expenses, customer information, staff, products, service, equipment, suppliers, and improvement ideas. Each Monday morning these Logs are faxed to Corporate Headquarters for senior executive review. In addition, any information (top-down, bottom-up and store-to-store) that would be of immediate use company-wide is broadcast as needed to all stores and headquarters via in-store fax machines. These avenues for communication ensure timely feedback and input from all levels of the organization as needed.

Through cross-training, each staff member has a complete understanding of all production and service procedures and quality standards to allow smooth transition from work station to work station and flexible response to volume cycles and unplanned reassignments. Work schedules and team assignments are developed by the Assistant Manager with heavy input from the Staff (Communication/Feedback Process). Process team members vary from shift to shift (three per day) and day to day, depending on individual availability and expected work requirements. Adjustments in assignments are made quickly when sickness, absenteeism and unexpected variations in customer counts occur. This emphasis on flexibility also results in decreased transaction times and consistent food quality results.

Pal's Business Excellence Process (including benchmarking, standardization, and continual improvement) is designed on the basis of piloting, studying, learning, and sharing learnings in both process team assignments and C-F Team assignments. Staff members (from hourly worker to senior executive) have continual learning ingrained into their basic approach to performing work. In addition, they participate regularly on improvement and learning teams to further incorporate continual learning into the culture and basic ways we think and behave. Brainstorming and consensus-building sessions enable these teams to identify problems or opportunities for improvements, analyze processes, and recommend solutions. This form of empowerment builds individual confidence, provides insight into problem solving, and leads to process and productivity improvements.

NEW PRODUCT, SERVICE, AND PRODUCT INTRODUCTIONS

Pal's processes are uniquely designed through its Business Excellence Process to interact with customers on an ongoing transaction-to-transaction, day-to-day basis. This ongoing interaction gives Pal's the opportunity to really get to know its customers and understand their needs. Products and services are carefully designed and product and service delivery processes continually measured and analyzed to ensure that operations are meeting customer requirements each and every time a customer is served.

Pal's is very careful never to introduce a new product or service without clear customer input, thorough market research, detailed design, extensive testing and analysis, and calculated rollout to ensure internal capability and customer acceptance. The New Product/Service/Process Introduction Process, featuring review and feedback from all stakeholders, is Pal's systematic approach for developing new or modified products, services, or processes.

The New Product/Service/Process Introduction Process begins with defining new product/service ideas directly from customer needs identified from customer listening strategies, market research, benchmarking, and communication/feedback with all stakeholders. After determining customer requirements, the Leadership Team considers new or revised product and service ideas, asking the following questions to perform an Initial Design and Value-Add Analysis:

Are our customer needs data and market trend data objective, valid, and credible?

- Is a specific product, service, or process suggested?
- How will the new product or service add value to all stakeholders?
- What new capabilities will Pal's need to do it?
- Will this change still add value to all stakeholders?
- What similar or related things are competitors doing?
- Do suppliers have the capacity to support this new offering?
- Is the appropriate technology available?
- Does Pal's have the capability to fully support this new offering?
- Will this change still add value to all stakeholders?

Initial Design and Value-Add Analyses of new product/service ideas are performed with a Desirability Feasibility Capability (DFC) approach focused not only on customer needs but also on operational requirements, both tempered with communication/feedback from all stakeholders. If the idea passes this first design phase, Pal's Strategic Planning Process is used to manage the Detailed Design and Piloting phases.

If initial DFC analysis indicates that an improved product, service, or process design will add stakeholder value, Pal's then uses Strategic Planning to plan the Detailed Design and Pilot Testing phases where Business Excellence Process is applied to research, design, pilot, and implement the improvement. Detailed Design and Pilot Testing are guided through the use of customer-focused process design criteria and new or modified product/service standards defined by the Leadership Team to ensure customer needs are met and process capabilities are established.

Pal's customer-focused process design criteria include requirements for

- Delivery temperatures
- Production times
- Portion control measures
- Changes in service times
- Vendor supplies and availability
- Storage
- Handling, and
- Health, safety, and sanitation

Pal's standards for new or modified products/services include

- Fit with current bread style
- Fit with existing store space/layout
- Consistency with existing taste profile
- Protein content used
- Capability and ease of staff training
- Potential for sales generation (at least 4% of Total Sales) and
- Potential for a favorable Unit Cost to Unit Revenue Ratio

During Detailed Design, Pal's applies benchmarking and the PDSA Cycle of the Continual Improvement Process to better define and refine the new product/service design idea. Input is obtained on supplier capabilities, available technology, and Pal's own capability to further refine the product and/or service design. Internal process standards and measures are defined based on established customer-focused criteria and proven operational cause-effect relationships. Benchmarking and the PDSA Cycle of the Continual Improvement Process are used to develop and refine the operational processes required to deliver the new product/service design idea. If possible, Pal's tries to simply adapt an existing process to establish the delivery capability it will need.

During the next phase, Pilot Testing, the new/adapted operational process is subjected to extensive piloting with a Standardize-Do-Study-Act Cycle (SDSA) approach to define/develop the final process design and capabilities. Process measures are evaluated vs. standards and targets to identify and remove barriers to standardization and projected capabilities. Potential new offerings are tested with three pilots using a hero store, antagonist store, and neutral store piloting approach to ensure that capabilities are established to meet customer-focused process design criteria and new or modified product/service standards. All pilot results and information are shared company-wide as data becomes available.

During the Pilot Testing phase, input is again sought from all stakeholders through the Communication/Feedback Process to make sure customer and operational requirements are satisfied by the new/redesigned product, service, and process. As a new product/service design is developed, suppliers are consulted to ensure they have the required supplies and quantities available to fulfill the quantities and varieties that our customers require. Pal's also determines if the supplier has any special requirements associated with this product or service and helps the supplier establish any additional capabilities needed.

Technology availability is also investigated. If new equipment is needed, Pal's makes sure during design that it will fit into the space available and that is it cost effective. Careful consideration is given to internal capabilities making sure that Pal's has the financial resources, human resources, and facilities to implement this product or service. In addition, the new product/service design is subjected to an initial exploratory test by focus groups brought together by the Creative Energy Group for in-store taste-testing of new or modified products.

Next, the new product/service design returns to the Leadership Team for extensive review and analysis using the preliminary piloting/testing data. If the design passes the Leadership Team review, it is then pilot-tested using a "Hero Store" (likes the design), an "Antagonist Store" (dislikes the design), and a "Neutral Store" (has no strong feeling either way). The results and data from this pilot test are evaluated by the Leadership Team to reach a consensus on a final plan for company-wide implementation. During implementation, process measures, MkBWA, and market research continue to ensure Pal's is fulfilling process and customer requirements.

If Pal's at any time during testing and piloting finds it cannot meet its criteria and standards for rolling out a new product or service, or if known customer requirements will not be met, the process is modified to meet requirements. If necessary, the new process design will be dropped altogether and a new design developed. Sometimes a failed design idea suggests that an improved product, service or process design still has the potential to add stakeholder value. In this case, Pal's then starts over with strategic planning re-applies its Business Excellence Process to research, design, pilot, design, and implement an improved idea that will meet the value-added opportunity.

The Leadership Team applies the New Product/Service/Process Introduction Process with extreme caution carefully calculating and planning new introductions to maximize impact, ensure training and standardization, and gain customer acceptance. Only after being thoroughly tested and after process capabilities are established and proven, does Pal's develop strategic plans for introduction of the new product/service with a store-to-store rollout. To avoid failures, lost customer loyalty, and tarnished image, Pal's never rushes a new product/service introduction or makes a menu change in a marketing panic. Its customer-focused, strategically-guided, systematic New Product/Service/Process Introduction Process provides the capability to introduce new products and services that are consistently characterized by coordinated processes and standardized quality within and across all stores, favorable customer/market response, and company-wide financial success.

PRODUCING EXCELLENCE THROUGH ON-LINE QUALITY CONTROL AND CONTINUAL IMPROVEMENT

Pal's processes for On-line Quality Control, Continual Improvement, and Benchmarking are all key process management components used to evaluate and improve core processes to achieve better performance, improved product/service designs, improved processes, and reduced cycle times. Through the On-line Quality Control Process, overall short-term results from automated control systems, personal on-line observations, and control charts are evaluated by process teams to identify improvement opportunities for core and specific product/service processes. Root cause analysis is used to pinpoint common barriers to consistent compliance with process standards; problem-solving is used to determine how to remove these barriers and reduce variability around established targets. Process flow-charting is also used to identify and evaluate inefficiencies or process steps that may be improved.

Use of the Continual Improvement Process to identify, prioritize, plan, and execute improvements is initiated both within individual stores by Process Teams and company-wide by the Leadership Team. Process Teams utilize overall long-term results from automated control systems, personal observations, and control charts to assess needs for improvement and appoint In-store Improvements Teams to apply the Plan-Do-Study-Act (PDSA) cycle and propose an improved product, service, or process design. Internal customer data are also gathered by Process Teams through constant day-to-day, on-line interaction among team members working side-by-side on the next step in the line and used to evaluate the need for improved process control and standardization. If any problems with a process or service are clearly identified, this information is shared company-wide via the Communication Logs.

The Leadership Team during monthly management reviews applies the Continual Improvement Process to analyze core and specific product/service processes then appoints cross-functional teams to address the needs or opportunity for improvement. The Leadership Team also analyzes the results of customer data from Market Research, Customer Surveys, and

MkBWA to identify needs for product/service/product improvement. In addition, recurring problems found in individual stores shared by Operators through the weekly Communication Log are routinely evaluated. These efforts are often done as part of the Strategic Planning Process and through our extensive piloting approach within the New Product/Service/Process Introduction Process.

New store openings are used as a special opportunity for company-wide process evaluation and improvement. Each Operator and a select staff member from each store work several shifts in the new operation and make notes of how the processes are operating. The observations are compiled, discussed by the Leadership Team, and used as input to the Business Excellence and Continual Improvement Process in search of new and innovative improvement opportunities. A constant emphasis for process improvement through these efforts by the Leadership Team and Process Teams is reduced cycle time to continually improve Pal's Sudden Service results.

MANAGEMENT REVIEW OF PROGRESS AND PERFORMANCE

The Leadership Team meets monthly to review data trends and variation, to assess performance toward goals, and to perform any problem-solving on the Business Excellence Process and leadership system that needs to take place. The Management Review Format used is outlined in Table 4. Key learnings are captured during these performance reviews and shared throughout the organization using the Communication/Feedback Process and the On-line Quality Control Process. The analyzed data from Management Reviews are then fed into the Business Excellence Process and used to assess the need for change in current directions and plans by

- Reassessing strategic goals
- Initiating any new benchmarking studies
- Consider new product/service/process ideas
- Pinpointing any improved process standards and targets
- Re-prioritizing improvement projects and
- Reallocating resources assigned to continual improvement activities based on new understandings of customer requirements and Pal's operational capabilities

Performance results and findings from monthly Management Reviews (e.g., identified root causes, improved procedures, control actions, and new learnings) are communicated company-wide (and as appropriate to suppliers) to keep performance on-track, to maintain standardization, and

Table 4   Management Review Format

Performance Area	Performance Reviewed	Gen. Staff Review	Asst. Mgr. Review	Operator Review	Ldrshp. Team Review
Financial Data	Sales	Daily	Daily	Daily	Monthly
	Profit			Monthly	Monthly
	Food Cost			Monthly	Monthly
	Cash Flow			Monthly	Monthly
Business Plan	Pro formas			Monthly	Annually
Competitive Data	Sales			Quarterly	Quarterly
	Turnover			Annually	Annually
	Health Inspections			Annually	Annually
	Prices			Quarterly	Quarterly
	Service Speed			Quarterly	Quarterly
Asset Productivity	Service/Labor Hr.	Hourly	Hourly	Hourly	Monthly
	Product Mix	Daily	Daily	Daily	Monthly
	Waste	Daily	Daily	Daily	Monthly
Operational Performance	Process Measures	Twice/Day	Twice/Day	Twice/Day	Monthly
	Sales	Daily	Daily	Daily	Weekly
	Turnover	Monthly	Monthly	Monthly	Monthly
	Profit			Monthly	Monthly
	Peer Reviews	Quarterly	Quarterly	Quarterly	Quarterly
	Staff Surveys	Annually	Annually	Annually	Annually
	Complaints	Daily	Daily	Daily	Monthly
	In-House Health Scores	Monthly	Monthly	Monthly	Monthly
	State Health Scores	Semi-annually	Semi-annually	Semi-annually	Semi-annually
	Service Speed	Quarterly	Quarterly	Quarterly	Quarterly
	Supplier Delivery	Per Occur.	Per Occur.	Per Occur.	Monthly
	Delivery Temps.	Per Occur.	Per Occur.	Per Occur.	Monthly

to facilitate organizational learning. When performance levels meet or exceed goal criteria, Pal's seeks new and meaningful ways to celebrate with its employees, customers, and other stakeholders.

Evaluations of these performance measures also help identify ever-improving ways to set direction, target improvement opportunities, implement Best Business Practices, empower the staff, and create a work environment that is successful in delivering excellence in the market served by Pal's quick-service restaurants.

CONCLUSION

The success at Pal's is evidenced through a review of its business results. In the important performance areas for Key Business Drivers (Quality, Service, Cleanliness, Value, People, Variety, and Speed), which Pal's has carefully linked to customer requirements, Pal's is consistently winning head-to-head competitions in its market against much larger national restaurant chain outlets.

Customer Satisfaction. Pal's most reliable indicators of customer satisfaction and dissatisfaction are Customer Transaction Counts, Customer Growth, and Complaints per 1000 customers, respectively. Measure of Customer Counts, which has shown steady growth at healthy rates for several years, reflects a positive overall performance trend for customer satisfaction, repeat business, and customer loyalty. The percentage of customer growth on a per store average also helps indicate the level of satisfaction of customers. In a market where competitors are showing negative customer growth Pal's has maintained a positive growth of additional customers. Customer growth for the first quarter of 1998 was 13.15%. Likewise, Pal's measure of customer dissatisfaction also reveals positive performance as its already low level of complaints shows a decreasing trend. Pal's level of complaints is far below the industry norm directly reflecting Pal's ability to understand the voices of its customers and the market and to address them with menu design, pricing strategy, sudden service performance, and customer contact standards.

Financial and Market. Customer Growth measure is also a principle indicator of Pal's financial and market success. As the number of customers doing business at Pal's increases, financial performance (Sales Volume and Profit) and market performance (Market Share) also improve. Sales Volume data present a healthy increasing trend over the past several years resulting from gaining new customers, increased business from existing customers, and successful new store openings. Pal's is approaching the leading competitor's (McDonald's) sales level in the market. The market leader has been on a sales decline for the past 2 years while Pal's continues to grow in sales.

In addition, Pal's Market Share continues to climb allowing it to pass major competitors and approaching the +20% level enjoyed by its leading competitor, McDonald's. Consequently, Pal's Profit Level data reflect an increasing trend that has resulted in effective supplier partnerships, effective process management, and improvement efforts. Pal's percent of profit from operation is above that of its best competitor and the industry norm. Pal's unique business strategy among QSR companies, which includes a customer-based focus on quality, service, and speed, has guaranteed its ongoing financial and market success and will continue to drive Pal's toward a business leadership position in its industry.

Human Resources. Staff Turnover, Pal's key quantitative measure of employee satisfaction, shows a highly competitive level of employee retention with an ever-improving trend. These results have produced all-time turnover performance lows for Pal's and are clearly superior to any competitor in our market. Furthermore, Pal's group of store operators and executives has enjoyed no turnover for the past 6 years, which has given Pal's a major advantage in leadership continuity over its competitors.

Employee well-being as measured by injuries due to Cuts and Falls also demonstrates an outstanding level of performance for Pal's. These reductions in Pal's largest types of injuries have resulted in a decrease in Worker's Compensation insurance rates from $4.20 to $2.75 per $100 of payroll. Employee satisfaction, employee well-being, and long-term (experienced) employees are key influencing factors on company performance in the areas of quality, speed, customer service, productivity, and customer satisfaction.

Staff Productivity is a Sales per Labor Hour measure that is Pal's primary overall indicator of Human Resource results. Pal's improving productivity trend reflects the effectiveness of its selection, hiring, training, and work designs, which are even further enhanced by our low turnover rate and more experienced staff.

Additional indicators of employee development and work system performance are the levels of participation in Pal's team structure and in its training program. All employees (100%) at Pal's serve on a least one and often two or more teams actively involved in managing and continually improving the business. Everyone is also involved in continually supporting personal, team and organizational learning and capability building. For example, Pal's key long-range human resource strategy to train 75% of its staff in ServSafe currently has 42% employees certified, far more than any of Pal's competitors. Pal's positive performance trends in customer satisfaction results, financial and market results, supplier and partner results, and company-specific results collectively demonstrate the positive results Pal's are realizing in human resource development and work system performance.

Suppliers and Partners. Through its Preferred Vendor Program, Pal's has consistently enjoyed excellent supplier performance from its four key suppliers. Pal's has successfully maintained long-term relationships with Coca-Cola for over 40 years, Institutional Jobbers for over 20 years, Lays

Packing for 15 years, and Earthgrains Bakery for over 2 years while achieving company objectives of no surprises, no delays, and no variability. Pal's Supplier Quality results reflect its high percentage of supplies that meet product specifications and arrive in sound condition.

Pal's Supplier On-Time Delivery results are consistently high despite tight scheduling of incoming materials to avoid out-of-stock situations, out-of-date product codes, spoilage, waste, and lost productivity. Delivery Temperature results for food products (per HACCP guidelines) are critical to achieving Pal's requirements for quality, taste, and texture, and avoiding bacterial growth. Pal's near-perfect results in these three critical supplier performance areas allow it to operate according to its tight requirements and consistently meet customer needs.

Operations. Food quality, customer service (accuracy and speed), food safety, and cost of operations are essential for meeting customer needs and financial requirements at Pal's. Food quality, which is built into the product designs and controlled by the product, support, and supplier processes, is confirmed by evaluations of Customer Satisfaction results. Customer Service results are evaluated in terms of accuracy (mistakes per 1000 orders) and speed (order cycle time). Order mistakes, which are a major source of customer dissatisfaction, have decreased by 10% over the past 3 years to a level of one mistake every 1000 orders (long-term goal = 1 per 2000 orders). Pal's order speed, which is a major determinant of customer delight, ranks it as the fastest restaurant in its market, twice as fast as the closest competitor (Taco Bell).

Food safety is evaluated by Health Inspection Scores, which is a primary indicator of Pal's compliance with sanitation requirements established by federal and state regulations and of Pal's ability to serve wholesome food products. Pal's consistently receives the highest Health Inspection Scores in its market and in the entire State of Tennessee.

Cost of operations has two major determinants, Food Cost and Labor Cost, which ultimately drive the profitability of a quick-service restaurant. Food Costs (expressed as a percentage of sales) are influenced by closely managed purchasing, tight inventory controls, strict operational standards, aggressive waste minimization efforts, and closely managed vendor relations. Food Costs at Pal's are continually declining and consistently at or below those of its closest competitor. Labor Costs are also evaluated in terms of Pal's Staff Productivity (Sales per Labor Hour) measure which doubles as an indicator of Human Resource results.

Pal's leadership role in its market has been achieved by producing dominant business results through satisfied customers, a capable and committed workforce, a supportive supplier network, and capable and efficient operations that deliver quality and speed at a reasonable cost.

Specific operational and service standards are established annually to drive Pal's to market leadership in various key performance areas. These standards and plans to generate the needed improvements will remain the primary topic at all planning sessions.

Pal's primary strategies, supporting goals and plans to achieve these performance projections include

- Keeping the speed of customer order processing time as a key driver for productivity and overall cycle times and an important consideration in Pal's planning process.
- Maintaining a pricing strategy for the next 2 years to hold Pal's value position of offering large portions of high quality food at its current prices. Pal's will provide larger portions and higher quality coupled with faster and more accurate service to define its competitive points of difference.
- Using the site selection process to help select profit centers to produce annual sales in excess of $1 million per year per new location. The number of stores will grow to at least 21 units over the next 3 years.
- Increasing Pal's market share to 20% over the next 3 years by having more locations that are convenient for the customers. As sales increase and food and paper cost decrease due to Pal's stronger purchasing position, profit margins will increase to 17% of sales.
- Increasing staff productivity with a pay-for-performance system that allows each person's pay to vary up or down on the basis of individual and company performance.
- Fully integrating HACCP and ServSafe throughout the company with 75% or more of all employees being certified. This will allow Pal's to focus on core process excellence while competitors are scrambling to meet what will by then be federal requirements for the restaurant industry.
- Capitalizing on available new technologies such as the Flashbake ovens or magnetic induction cooking that will be affordable to install to reduce product processing and preparation times. The other major chains will also have this technology available but, as with microwaves, Pal's will be the first to make use of it in its market.

The Business Excellence Process will continue to drive Pal's toward continual performance improvement over the next 2 to 5 years through strategies and plans addressing meaningful values, high performance expectations, staff empowerment, a keen focus on customers and other stakeholders, learning, and innovation.

COMPANY

Ulster Carpet Mills Ltd.

PRODUCT/SERVICE:

Manufacturer of woven Axminster and Wilton carpets

LOCATION:

Portadown, Northern Ireland and Durban, South Africa

NUMBER OF EMPLOYEES:

1200

KEY PERSONNEL:

Case Study Contributor: Daniel McLarnon

KEY USES OF INFORMATION:

- Company-wide self assessment
- Quality award feedback
- Customer feedback (surveys, face-to-face interviews, customer facing groups)
- Process reviews

- Supplier feedback
- Employee feedback
- Benchmarking studies
- Competitor analysis information
- Social issue information

ORGANIZING FRAMEWORK:

1. Production and Technical
2. Sales, Design, and Administration
3. General

AWARDS AND SIGNIFICANT RESULTS:

- Winners of Northern Ireland Quality Award
- Winners of United Kingdom Quality Award for Business Excellence
- Winners of United Kingdom Safety Award
- Exports up 700%
  - Output per employee up 60%
  - Sales up more than 130%

INTRODUCTION AND BUSINESS OVERVIEW

Ulster Carpet Mills is a family-owned, professionally managed manufacturer of woven Axminster and Wilton carpets, established in 1938. Today the company employs some 1,200 across two manufacturing plants and a world-wide sales, marketing, and design network. Manufacturing sites are in Portadown, Northern Ireland and in Durban, South Africa.

The group also embraces Ulster Carpet Mills (N. America) Inc., whose head office is in Atlanta, Georgia.

Ulster Carpet Mills began its Total Quality journey back in 1983, when a number of Quality Circles were introduced throughout the company to improve efficiency and operating performance.

The company achieved ISO 9001 Registration in 1986, and developed a company-wide Total Quality process in 1989, starting with the first program, which it called "Total Customer Satisfaction." In 1992 the process was reviewed, training was again carried out from board levels to shopfloor and across all administration and sales functions and relaunched into a second program called "Total Customer Delight."

Company-wide Self-Assessment using the European Business Excellence Model, which is based on the Malcolm Baldrige Award model, was introduced in 1993.

In 1995 the company launched its third program on its Total Quality journey called "Bridging the Gap," i.e., bridging the gap from where it perceived itself to be under self-assessment to where it ideally would like to get to in terms of customer delight, employee delight, and business performance. The company's achievements in Total Quality were recognized when it was awarded the Northern Ireland Quality Award in 1991 and the UK Quality Award for Business Excellence in 1996. HRH Princess Anne presented the latter award at a prestigious dinner evening in London.

"SNAPSHOTS" OF THE JOURNEY

In the Beginning...

When the program "Total Customer Satisfaction" was introduced, it was apparent to Ulster Carpets, at that stage, that around 30% of employees were open to change and the need for change. Another 10% were deemed eager volunteers. On the other end of the scale about 10% of employees were classed as skeptics. Ulster Carpets' strategy at the beginning, therefore, was not to try and convert the skeptics, but to pull as many of the remaining 60% of the workforce toward the "converts" side.

If after the initial 2 years the company had grown its 30% willing participants to become 60% or 70%, then it would have a very substantial number of its people involved in continuous improvement activities. This, it was felt, would then apply peer pressure to others, thus creating a knock-on domino effect.

Building the Platform

In order to understand where we were as a company, a consulting group was employed to carry out a "Diagnostic Survey" of the company with its customers, suppliers, and employees. A presentation to the Board on the findings from the survey was made and from this it was agreed to have a series of off-site workshops. The participants at these workshops included all the senior management team, all the middle management team, a cross section of supervisors, all the sales force, and all the union representatives. In addition to presentations explaining the philosophies of Total Quality, Teamwork, and Customer Satisfaction, the participants were asked, through syndicate exercises, to outline the strengths of Ulster Carpets, what they felt were the weaknesses and what they saw as the obstacles and fears to making Total Quality a way of life within the company.

The outputs from those early workshops included 180 "suggestions for improvement" from the participants. In reality, the "suggestions" were finger-pointings. "If only that was sorted out for me or my team by those other people, how much easier my life would be—it is all their fault."

And this was the main objective of the first Total Quality program: to lead people in turning the fingers around to point to themselves saying, "This is what I and my team need to do for my internal customer, to improve our service or product."

To start the ball rolling the 180 "opportunities for improvement" were categorized into three project groups: "Production & Technical," "Sales, Design and Administration," and "General." Within each group the opportunities were further categorized into short-, medium-, and long-term projects. The short-term projects were sorted out quickly with plenty of fanfare and communication. In parallel to this happening, the senior management team went through a 3-day training course, again off-site, which covered all aspects of Total Quality in much more detail.

Five of these senior managers were selected and trained for a further 5 days, and it was they who presented the material to the middle management and supervisory levels within the company.

So now Ulster Carpets had five senior people who, over 3 days, could stand on their feet and present Total Quality issues covering customer needs and expectations, problem-solving tools and techniques, and teamwork, etc. Brainstorming and cause-and-effect exercises were carried out using "live" themes provided from the "opportunities for improvement" lists of the first workshops.

Having then trained all the management team from the Board on down to supervisor level, Ulster Carpets then put all these people into teams to work on various problems. This meant that the company chairman, the managing director, directors, middle managers and supervisors were involved in either leading or being members of problem solving groups. And this was all before any shopfloor training or involvement in the Total Quality process. The purpose was to gain commitment from managers at all levels and to demonstrate to them and to the rest of the workforce that this thing called "Total Customer Satisfaction" was something special.

After initial training and with management-involved projects underway, it was time to turn attention to interdepartmental relationships. The aim was to get departments to understand the needs of their internal customers, an approach based upon having formal supplier customer contracts.

Workshops were arranged at which customer/supplier departmental managers were brought together to identify three things that the supplier department would agree to improve for its next-in-line customer department: three things that would make a significant improvement. The customer department measured the current performances of the things to be improved; department members agreed on improvement targets for the three improvement areas and timings by when the improvements should be made.

The two managers then signed the formal contracts, "in blood." The contracts were then framed and hung in the customer department, with performance charts placed alongside. Performance to targets were recorded and marked up by the customer department. These contracts were set up for all departments—operations, sales, administration, R&D, etc. It was approximately 6 months and much communication later that attention was turned to those levels within the organization below supervisor level.

Seventeen managers and supervisors across all functions were identified and trained to carry the Total Quality message to the rest of the workforce. These managers carried out their training over 3 days and therefore had to be able to convince their direct reports of their commitment to the process and teach employees about problem-solving tools and techniques.

Total Quality knowledge had by this stage grown to a high level. Many more project improvement teams were put into place across all levels and divisions. The fingers of responsibility for improvement were starting to turn away from "what should be done for me" to "what I and my team can do for our customers."

Team Building

To emphasize the strengths and advantages of teamwork Ulster Carpets organized off-site, team-building weekends for everyone within the organization. At these sessions, which were fun weekends, natural work teams went through a series of Outward-Bound-Style adventure projects. Planning, problem-solving, and effective communication were essential ingredients for successful completion of the exercises. The only restraint imposed on the team members was that they each had to lead a part of each session. The advantages of teamwork and team knowledge were brought to the fore during these sessions and as a spin-off, stronger bonds were made within each group.

Telling the Story

Communication increasingly has become a powerful weapon in the armory of change management. Deliberately, Ulster Carpets has measured all key processes and communicated their performance company-wide. The company photographs people both individually and in teams, and their project improvements. Huge, brightly colored communications boards are fitted with spotlights. When something new goes on a board a spotlight is shone onto it to highlight it. All employees are briefed every month. The company has not missed a month's briefing since it started in 1992. Today, Ulster Carpets tells its employees everything that happens in the business: sales levels, sales revenue, credit-note payments, debtor-days, profit levels and margins, new orders won, production levels, orders running late, orders of highest priority. It also tells them births, deaths, and marriages relating to employees and their family members, and about visitors to the plant, thank-you letters from customers, etc.

All questions raised across all briefings are recorded—even though they may have been answered at the time—and placed on all the noticeboards across the organization so that everyone can see what questions were raised at each briefing session. Important news that breaks in between monthly briefings is communicated through an alert system. News is either distributed via a Red Alert (so called because the message is typed onto red notepaper) and communicated to all employees immediately as it is received or sent out as a Yellow Alert (on yellow notepaper) and communicated at the first opportunity, i.e., at a break or start/end of a shift. Neon moving-message boards are used to flash up information—from a birthday greeting to names of visitors who may be touring the plant. Company Purpose and Values statements are written onto large boards and hung from the ceilings around the plant as reminders and to gauge adherence to them.

POLICY AND STRATEGY

How Policy and Strategy Are Based on the Concept of Total Quality

The company mission statement was drawn up in 1989 by a team representing all levels and covering the main functions within Ulster Carpets. The team was led by the then Chairman, Edward Wilson. A review team, again led by Edward Wilson, reexamined the Mission Statement in 1992 when Ulster Carpets was further drawn down the total quality route to ensure that the statement covered all aspects of its business and direction. The words "delighting our customers" replaced "satisfying our customers" and Ulster Carpets' vision for the environment was added.

The change of emphasis from "satisfying" to "delighting" is significant as this is central to the formulation of company policy, strategy and subsequently business plans. Values statements are displayed throughout the organization and Ulster Carpets uses its triannual magazine, monthly briefing, etc., to keep emphasizing them as they are core to company strategy and link directly into its Mission. This has been pivotal to Ulster Carpets' drive and enthusiasm for continuous improvement since 1989. The company recognized that it needed to establish a quality culture which would permeate throughout the whole organization. The direction it took was set through its 5-year Forward Development Plan (FDP), which is assessed monthly and reviewed annually.

To fulfill its Mission and keep its values to the fore Ulster Carpets needed to

- Change Total Quality emphasis about every 2–3 years
- Ensure communications are meaningful and colorful
- Drive total quality process through the senior management group
- Always put the customer first

This has prompted the company to move its Total Quality programs from "Total Customer Satisfaction" to "Total Customer Delight" to "Bridging the Gap." Since 1989 three Total Quality Coordinators have stamped their own enthusiasm into supporting the venture. In April 1994 the company restructured into four Business Teams as a result of feedback from customers and employees via review meetings, strategy meetings and surveys, together with its own analyses of the marketplace and competitors. Two of the Business Teams are focused onto the two distinct external customer groups, both of which have mutual as well as individual needs and demands.

Strategies

Strategies are set from the Mission and Value Statements and are translated into action plans targeted specifically at Customer groups. Strategies and action plans for each facet of business are set to

1. Ensure that customer needs are understood and where possible surpassed
2. Provide a product and a service that delights customers and is the envy of competitors
3. Provide manufacturing technology that provides competitive advantage and anticipates customer needs
4. Develop and train employees to Ulster Carpets' needs and vision
5. Maximize flexibility through development of group working
6. Provide on-time delivery, as agreed with customers
7. Ensure the continued financial viability of Ulster Carpets

How Policy and Strategy Are Based on Information That Is Relevant and Comprehensive

Customer Feedback

Feedback from customers is obtained through

- Retail customer surveys:
  - Phone survey the top 150 customers
  - Postal survey all others
- Face-to-face formal interviews with visiting customers
- Trade exhibitions and shows for feedback on
  - New design ranges
  - Delivery and service
  - Sales support
- Customer facing groups within Ulster Carpets:
  - Sales
  - Designers
  - Installation planners
  - Installers

Information received that requires immediate action is fast-tracked into the system. Short, medium, and long-term needs are positioned accordingly as strategy is fed through to form part of future policy. Examples include the setting up of global sales and marketing subsidiaries; the positioning of designers and planners worldwide; developing new, plain background carpet designs; developing a partnership with a single transport company; and introducing a unique 60-day "no-quibble" guarantee on residential carpets.

Project teams were set up to advise on how to effectively implement some customer feedback requirements. The need to reduce dispatch times ex-stock from 2 weeks to 24 hours and answer the phone to external customers within three rings are examples of project team improvements to customer needs.

Supplier Feedback

In 1990 Ulster Carpets reduced the number of main suppliers and formed working partnerships with those it chose. This stemmed from the Mission statement *"Our suppliers: Working with us in delighting our Customers."* Suppliers recognize benefits to both parties on a win–win basis. Ulster Carpets obtains supplier feedback from

- Monthly review meetings with its transport supplier
- Visits to the factory by Ulster Carpets' material supplier representatives and senior managers
- Supplier training days at Ulster Carpet Mills
- Supply contract review meetings
- Visits to suppliers by Ulster Carpets' quality personnel, production managers, and directors
- New technology and new machine installations

Feedback from suppliers has improved

- Ulster Carpets' administration to them
- how Ulster Carpets' identifies its products
- carpet roll packaging for export
- suppliers' appraisal of incoming goods
- delivery time to customers
- suppliers' planning ability through better order control by Ulster Carpets

Employee Feedback

The company value relating to Ulster Carpets' people states "We value our employees" and as the company recognizes that total quality is all about people, it looks upon feedback from its employees as essential to the success of its business. Ulster Carpets obtains employee feedback formally through

- Employee opinion surveys
- Project team presentations
- Personal development plans
- Two-way briefing sessions
- Safety meetings
- Business Team Safety group
- Central Safety group
- Syndicate exercises at training sessions
- Face-to-face self-assessment interviews
- In-house annual sales conferences
- Investing in People interviews

Employee feedback has impacted greatly on Ulster Carpets' quest for delighting both its internal and external customers and helping its suppliers.

Benchmarking Studies

All Ulster Carpets' 11 senior managers, including the Chairman and Managing Director, made weeklong best practice study tours to the U.S. with the N. Ireland Quality Centre to visit companies that had won the Malcolm Baldrige National Quality Award. Each study tour had a 1-day planning session so that each participant targeted specific areas to address such as Communication, Teamworking, Customer Service, etc.

Ulster Carpets' Chairman and four of its managers attended week-long best practice study tours to British Quality Award Winners, organized separately by the quality group of the Textile Institute and the N. Ireland Quality Centre.

All eight manufacturing managers made benchmarking trips to Akeda Hoover, which makes car seats for Nissan, on improving delivery performance and to Amdahl on problem-solving techniques.

Ulster Carpets is a member of a N. Ireland benchmarking group and continually collects information on competitor performance for comparison with best in class.

Competitor Analysis Information

Ulster Carpets collects competitor data from

- Monthly sales and export information from the British Carpet Manufacturers Association
- Trade magazines throughout Europe
- Company reports
- *Financial Times* and *The European* newspapers
- Performance Ratio Analysis Reports

Information collected is used as part of company policy and strategy review and to help identify and set competitive strategy for the main European market areas. Market trend information is circulated to all directors and senior managers. Sales and Marketing personnel use the information to identify opportunities and possible new design ranges to keep Ulster Carpets the recognized leader in product design and color. Competitor analysis also allows Ulster Carpets to compare technological developments against its own strategy team. Lack of thrust by competitors in loom speeds prompted Ulster Carpets to invest in developing high speed looms which will greatly improve delivery times as well as provide a distinct price advantage. Based on feedback Ulster Carpets is also pioneering revolutionary yarn dyeing technology in association with an Italian machine manufacturer and a German dyestuffs group.

Social Issues

Ulster Carpets also collects information for comparative purposes on

- Energy use
- Prevention of pollution
- Regional and national wage rates
- Community requests for visits

This information has led the company to drill a bore-hole during the summer of 1995. The main manufacturing site is 85% self-sufficient for water. Ulster Carpets works with the N. Ireland Electricity Service to run its own generators during peak winter power requirements.

Ulster Carpets is also one of the top local wage payers and within the top 5% of companies in N. Ireland for wage packages.

Auditing Our Systems

Ulster Carpets is accredited to ISO 9001 and is in the final stages of implementing ISO 14001, environmental standard. The company also has obligations under health and safety legislation. In order to ensure that the procedures are integrated within the normal working practices of the business and to further strengthen employee knowledge Ulster Carpets has devised auditing procedures that are "friendly" to allow work teams to carry out assessments.

Each month each Group Leader selects two members of the team who carry out procedure audits of their own work area against the two standards. Each Area Safety Committee carry out monthly safety audits against key safety criteria. The findings from all the audits are scored as Gold, Silver, or Bronze, depending upon their compliance with good practice. Where an item is outside tolerance it is given "Red." The teams are empowered and responsible for ensuring that any problems found are corrected.

How Policy and Strategy Are the Basis for Business Plans

The development and growth within Ulster Carpets has been due in many ways to its approach to business planning. Ulster Carpets has moved over the last 5 years from a "home player" into a global organization. Steps along the way have been carefully planned and communicated with strategies set in place. The deployment always starts with a review of values and vision, comparing the company it is to the company it wishes to become. A review of the previous year's performance against strategy is assessed and business considerations are then put forward for the following year. All areas of the business are considered at this stage: sales projections, marketing needs, investments, global issues, competitor performance, etc. It is also at this stage that customer and employee inputs are added. This first phase usually takes up a full day in review and analysis.

Senior managers then set out their strategies for their areas. These strategies cover, where appropriate, sales projections, production levels, capital expenditure needs, total quality, budget proposals, people development, safety and even a "wish list." Together with scenario-planning the outline strategies are agreed, but each must show how the elements fit the mission statement and company policy list.

The final plans, which may have taken a number of drafts to finalize, are briefed by the Senior Managers company wide. They also form the backdrop of the Managing Director's "State-of-the-Nation" address. Employee feedback is encouraged and fed into the business team action plans set to ensure the strategy happens. Functional objectives are then set from the action plans. The monthly Senior Management Group meetings, which typically run from 10 A.M. to 3 P.M. (with working lunch), review the plan and adjust where necessary.

Any changes and performance to plan are briefed monthly companywide. The whole business planning cycle runs from November of one year to February of the next.

How Policy and Strategy Are Communicated

The policy deployment process is seen as the most efficient and effective route for communication of the company's policy and strategy and derived business plans. As with training, policy and strategy are cascaded down through the organization by the following:

- Presentation and discussion with managers across all functions
- Prepared summaries, using overhead transparencies to give a common message, are presented to all other employees face-to-face. Discussion is encouraged.
- Company overview presented by the Managing Director to all employees in May, with question/answer session following
- Company magazine carries keynote address by Managing Director together with articles on aspects of current strategies
- Monthly company-wide briefings keep employees up-to-date and give an opportunity to question and feedback views. All questions raised and answers given are published on the noticeboards.

In order to keep policy and values to the fore key statements are displayed throughout the organization, so that employees can compare strategies and direction against these. Important information that needs to be presented to employees is communicated through the "Yellow" and "Red" alert system. Effectiveness of communication and the deployment are assessed through

- Briefing session feedback (numbers and types of questions asked)
- Employee opinion surveys
- Directors and managers talking to employees
- Comments made during the face-to-face and one-on-one self-assessment interviews
- Annual sales conference, which brings together the sales team, designers (including Kidderminster), and London personnel

How Policy and Strategy Are Regularly Updated and Improved

Ulster Carpets has been using a policy and strategy review mechanism since the start of its company-wide Total Quality process in 1989. It is essential that Ulster Carpets' policies, values and strategies link into one another. Dedication to seeking to delight customers and the involvement of Ulster Carpets' people in continuous improvement are the bedrock of the company's quality policy and strategy.

Ulster Carpets sets strategies that give direction yet are flexible enough to respond to customer needs and marketplace changes. Each month at the Senior Management meeting the company systematically reviews the effectiveness of its strategies from both measurements and comments within the organization and views and analysis external to the organization. The Senior Management Group consists of the Chairman, Managing Director, Company Secretary, Directors of the four Business Teams, Sales Director, Research and Development Manager and the head of the recently set up Rug and Sample Division.

The first item on the agenda is Total Quality performance and review, even before the financial assessment review. Strategic items cover the rest of the meeting.

Each Senior Manager plans against set targets and milestones. Presentations focus on areas such as

- "Hard" performance measures: sales, production volumes, cycle times, waste levels
- Customer Delight: on-time delivery, customer thank-you letters

■ Employee involvement: number of project teams, number of people involved in improvements, formal recognitions

Therefore, the reviews give an insight into which strategies are on target and which require action or support. These monthly reviews give virtual real-time updates and constant improvement as and when necessary.

Each quarter and year end, the same presentation format is presented to the Holdings Board of the Ulster Carpet Mills group. The quarterly reviews ensure that the strategic plans are maintained and any adjustment due to market changes, legislation, or competitive threat is made quickly. Similarly, if an in-house Research and Development breakthrough pulls part of the plan forward, then this can be discussed and funded, if necessary.

Problem-Solving

People involvement in problem-solving was partially led and partially driven in the early years. Over time, involvement has become self-generating, as barriers have been broken and fear eroded. Ulster Carpets now has around 300 teams in action per year covering all aspects of the business.

A key problem-solving aid has been the use of CEDAC (Cause and Effect Diagram with the Addition of Cards) boards. Each shows the large fishbone diagram with white cards detailing the problem under investigation placed at the "head of the fish." Brainstormed suggestions of possible causes are then placed around each "rib of the fish," and ideas for solution positioned opposite each probable cause. In office areas, where space is restricted, smaller CEDAC boards and "post-it" notes are used. In both cases the CEDAC boards are portable so that the teams can take them into the project rooms to work on the problems.

How Ulster Carpets' Processes Are Reviewed and Targets Are Set for Improvement

For Ulster Carpets, continuous improvement and change are everyday occurrences and part of the fabric of the organization. The standard review process throughout the organization ensures continual identification of areas of improvement opportunity. Feedback is derived from a number of sources.

External and Internal Diagnostics: These interviews gave direct customer feedback on their perception of Ulster Carpets. The company then focuses on building on its strengths and reinforcing areas of weakness, for example.

Telephone Response Times: This was an area of particular concern for customers in contacting the main switchboard. The company invested in a computer software package that monitors all phone calls. The target response time for incoming calls at the switchboard is 5 seconds and times are charted. If this response time rises above 5 seconds it is due in the main to responding to extended customer requests.

Stock Orders: In 1989 the cutting and dispatch process initially took up to 10 working days to cut and dispatch carpet from stock. In 1991, 90% of stock orders were cut and dispatched within 24 hours. Currently 99.9% are dispatched within 24 hours and 100% within 48 hours. This has been during a period where sales have increased by 48%.

Internal Communication Problems: New company structure has greatly aided communication.

Dyehouse and Yarn Supply Problem: Delay in supplying yarn on time to looms impacted upon lead times. Average downtime is currently 1% against a high of 13% in 1993.

Customer Surveys: The major driver behind the changes in stock carpet patterns from traditional heavily patterned to plain background designs.

Employee Surveys: The company has made a number of changes resulting from employee feedback.

Complaints Analysis

On-Site Customer Interviews

Benchmarking Visits: Some examples of benchmarking visits are

Akeda Hoover—Logistics and delivery process
Dupont—Health and safety processes
Amdahl—Process improvement techniques
Bombardier Shorts—Wages process
Harland and Wolff—Total Preventative Maintenance process

The Company is an active member of the Northern Ireland Benchmarking Group, which meets twice annually to compare best-practice.

Internal Benchmarking: Comparisons of output within the group.

Quality Audit Reports: Monthly internal departmental audits.

Internal Accounting Ratios: The company is very focused on control of stock levels and work-in-progress. During the past 5 years the company has grown its portfolio of design products but has managed to reduce the finished goods stock level and raw material stocks without any decline in service levels to the customer. This has been facilitated greatly by the introduction of a forecasting system **LOGOL** and the link it plays with new production planning systems. A manager has been appointed to keep complete control over this whole area.

Management Accounting: Each individual process owner now has direct responsibility for purchasing and cash flow management. The monetary targets are set and reviewed within business team meetings and at the monthly Senior Management meeting.

New Activity Based Costing systems (ABC) are utilized for costing products prior to product sales. The company employed a full-time Management Cost Accountant in 1994 who has responsibility for these systems. This has been crucial for the Company in the market place for pin-pointing pricing strategies.

Targets are set for processes and for individuals by

1. Customer feedback
2. Machine capabilities
3. Agreed personal objectives within the Personal Development Process
4. Historical data
5. Internal contracts

Target setting is reviewed annually, unless a more immediate change is required, e.g., sample production had to increase capacity in the middle of year 1994/95 to meet increased, unexpected stock sales.

Process Reviews

All Ulster Carpets processes are monitored regularly to ensure that targets are being met. The weaving process is a perfect example of how the company has developed its own method on each loom of recording production rates efficiencies and downtimes automatically and continually. This system is known as data-logging and has given the company a great advantage in production planning. Examples of reviews are

1. Weekly Quality Reports
2. Quality reviews
3. Complaint Analysis
4. Internal Benchmarking and management indices
5. Data-logging reports
6. Monthly Delivery Cost and Time Analysis

These reports are information sources that facilitate immediate action on areas of deviation. If immediate action and improvement are not successful then the process team investigates the area in order to review the process and to make recommendations for improvement.

How Ulster Carpets Implements Process Changes and Evaluates the Benefits

Proposed process changes are carefully evaluated and if any funding is necessary it is assessed and actioned. The process owner and the team are responsible for the changes and improvements, monitoring their effectiveness as the changes proceed. For any system changes, the Quality System Auditor gets involved to ensure that all quality manuals are updated and the improvements are documented effectively.

How Staff Are Trained Prior to Implementation

All training requirements of employees are planned by training department personnel and the process owners. Training needs are identified through the annual Personal Development Plan reviews at which a documented training plan is developed. If changes to the process require special training of personnel, then often the team members themselves are involved in this, e.g., for the new customer services computer system, the customer service staff were involved in developing the system alongside computer professionals and trained so they in turn could train others.

If new machinery is being purchased, the training department and the process owner are involved in the purchasing agreements, in order to ensure skilled and safe training is provided by the supplying company.

The effectiveness of training approaches are regularly reviewed, particularly during the piloting stages of major changes.

How Process Changes Are Communicated

Process changes are briefed company-wide through the Team Brief process where possible, and through departmental presentations when necessary, so that all relevant employees and internal customer and supplier groups are informed of the impact the change may have on their own job functions. This also provides direct feedback from individuals and teams involved with the change. A variety of communication channels is used to communicate planned changes to processes. The company targets and tailors messages for all levels involved or affected by the changes.

How Process Changes Are Reviewed to Ensure Predicted Results Are Achieved

New performance standards relevant to planned changes are agreed and reviewed by the process owner and Business Team Director. Targets are agreed for the improvement, and subsequent reviews monitor the impact of the change and if other influences may be affecting the change. Processes are charted in each department, so all staff can see any part of their process under review, or process changes made.

Post-implementation reviews also allow the company to understand the improvements the change has made or problems that may have arisen. The impact of Ulster Carpets' approach to change management manifests itself in its business results.

Process Driving Results

The key process for taking all organizational information and converting it into business results has six broad steps:

1. Planning/forward development plan (scope, comparatives and data)
2. Pilot scheme (training and process owner identified)
3. Analysis (performance gap and future levels)
4. Integration (Internal customer-supplier assessment)
5. Implementation (action plan, team playing, company brief), and
6. Review

The analysis and the integration steps give Ulster Carpets key knowledge upon which to base implementation decisions.

Impact on Business

The success of Ulster Carpets' key and critical processes is demonstrated through customer satisfaction measures. The customer can tell the Company directly in terms of customer delight performance and business results measures whether success has been achieved. These measures are derived from assessment of the needs of customers and Ulster Carpets' employees.

The other feedback methods utilized to help facilitate and identify areas for future improvement and to develop effective products and customer services include customer surveys, interviews, phone surveys and independent consultant interviews.

There are also three people dedicated to customer services acting as direct feedback from the market place.

The Company has illustrated commitment by investing heavily in information technology systems to record market information. Ulster Carpets measures the performance of its carpet delivery and customer support processes across a range of quality service criteria. For business processes, the information provided from these measures helps the Company assess its competitive standing within its target markets.

All of the monthly quality costs are directly allocated to processes and departments within three yearly trends. This information is distributed to all levels. Business Team Directors and process owners discuss areas of non-conformance and take appropriate action.

Achievements and Accolades

- All employees, from the company chairman down to the shopfloor, have been involved in project improvements.
- At any one time around 30% of employees are actively involved in continuous improvement project activities.
- Savings per project have ranged from £0 to £0.5 million
- Winners of N. Ireland Quality Award
- Winners of U.K. Quality Award for Business Excellence
- Exports up 700%
- Output per employee up 60%
- Moved from world number 4 to world number 2 Axminster carpet producer
- Sales up more than 130%
- Profitability held in a very competitive market

8

HOW TWO WORLD-CLASS PRIVATE SERVICE COMPANIES IMPLEMENT PRINCIPLES OF PDKA— BI PERFORMANCE SERVICES AND BRITISH TELECOM NORTHERN IRELAND

INTRODUCTION

The companies that provided these case studies—BI Performance Services and British Telecom Northern Ireland—offer distinctly different services. Both have won many awards, and both implement several core principles of the PDKA process.

BI Performance Services

BI Performance Services (BI) is tied directly to the performance improvement industry: BI combines communications, training, measurement, and rewards into performance improvement programs for its corporate customers nationwide. BI is headquartered in Edina, Minnesota, a suburb of Minneapolis, with 27 offices throughout the United States. BI's mission is simply to improve its customers' performance. BI achieves its mission in a variety of ways: improving market share, cost reduction, introducing new products, developing brand loyalty, generating ideas from the employee base to improve productivity, measuring customer satisfaction, recognizing employees, and developing customer loyalty. Customers are primarily buyers from Fortune 1,000 companies who outsource performance improvement programs. Creativity and innovation are key to providing value to BI's customers and in fact are how BI attracts and retains customers; both are customer requirements.

BI has a very mature customer interaction and survey methodology. Its Transactional Customer Satisfaction Index (TCSI) process follows the logic of PDKA. Three major factors—pervasive customer focus, strong Associate focus, and the BI Way—are the cornerstones of BI's success. More than 75% of BI's Associates are regularly involved in customer contact. With services custom designed to meet individual customer needs, and long-term relationships with customers, there is intense customer focus throughout the organization. Such approaches as the relationship management and sales processes, the Customer Delight Process for service design and delivery, and the Transactional Customer Satisfaction Index process embody this customer focus.

In 1994, BI became the first service company to earn the Minnesota Quality Award. BI has also been awarded GTE Directories Supplier Quality Award, GM Supplier of the Year, and Ford Marketing Excellence Award. In 1998, BI was selected as the only service company nationwide to receive a Malcolm Baldrige site visit, which is the final step of the award process.

British Telecom Northern Ireland

British Telecom Northern Ireland (BTNI) is located in Belfast, and its primary service is telecommunications. British Telecom Northern Ireland has had significant findings in terms of obtaining leadership buy-in and acceptance of feedback, and of how the lack of acceptance can stifle further organizational improvement. BTNI has obtained ISO 9001 certification and has won both the Northern Ireland Quality Award and the United Kingdom Quality Award.

HOW THESE COMPANIES HAVE IMPLEMENTED THE PRINCIPLES OF PDKA

Both of these case studies have identified and addressed a single key information source for their businesses and have explained how their specific information source is systematically using the PDKA logic. BI has focused on its TCSI customer feedback process and BT Northern Ireland has chosen to focus on its quality style assessment feedback reports for the purposes of their case study. Also, BT Northern Ireland addresses an interesting perspective on the criticality of obtaining leadership buy-in and commitment.

Executive Buy-in

BTNI believes that an assessment will result in an objective feedback report clearly identifying the applicant's strengths and areas for improvement (AFI). It is generally assumed that the applicant will use the feedback to identify the systematic factors and process weaknesses that are responsible for poor performance. However, BTNI contends that the initial response of the applicant's senior management team to the feedback report is crucial for the future success of self-assessment within the organization. A key point is that personal commitment of managers can lead to emotional reactions to external feedback and may present a barrier to the ongoing success of self-assessment and involvement. BTNI believes that based on emotional reactions, many opportunities for improvement from feedback reports may be ignored. To overcome this barrier, BTNI proposes that management employ a transition phase model to assist them in coming to terms with the initial feedback report and in their eventual implementation of needed improvements. BTNI's experience of the transition phase model recognizes that failure to accept the external feedback reflects misgivings in the rigor of the entire self-assessment process, and follow-up exercises are unlikely to be undertaken. Those companies that do progress, on the other hand, have learned much from their first exercise. In the case of BTNI, the company integrated its second feedback report. This reflected an acknowledgment of the complete self-assessment cycle and the need to act on the feedback in order to advance.

Action Plan Teams

BI's measures from the strategic measurement and analysis model are aggregated and analyzed by a variety of action teams. Action teams are tasked with understanding the feedback from the customer. An action plan is created to take the survey responses from the TCSI and clearly understand the feedback so that the information can be used effectively.

Consolidation of Information

BI's key uses of information are self-assessment, Malcolm Baldrige Award feedback, and customer input. BI gathers a variety of information types;

however, for its case study, it has focused on its key information source, which is customer feedback. BI has a Transactional Customer Satisfaction Index (TCSI), which is an actionable survey that is both transactional and designed to gather, integrate, and implement customer input. The TCSI process is used to determine both customer requirements and satisfaction with key transactions. The TCSI process integrates two steps—expectations management and transactional measurement—to strengthen customer relationships and improve customer satisfaction. The TCSI process has two survey components: the customer survey and the delivery readiness survey.

The customer survey process steps are to (1) complete the TCSI form on line; (2) work with the client to determine and hold Expectations Discussion meeting; (3) enter survey responses into the system; (4) develop and sort quarterly reports; and (5) review and analyze results with teams.

The Delivery Readiness Survey process steps are to (1) customize the survey; (2) complete the Delivery Readiness Survey; (3) collect responses and review for concerns; and (4) fix any concerns. The TCSI process includes four phases: delivery team meeting where they prepare to talk with the customer; customer expectations discussions where BI clarifies customer expectations about what it provides to the customer and how; surveying the customer after BI completes major phases of delivery; and using survey results from all aspects of its business for action planning.

BTNI, like other companies, has a variety of information sources; however, it has chosen to focus on its quality style assessment feedback reports for the purposes of its case study. BTNI's key information revolves around external feedback reports such as ISO 9001 audits, British Telecom Award feedback, Northern Ireland Quality Award feedback, and United Kingdom Quality Award feedback. It also gathers self-assessment feedback.

BTNI believes that integration of assessment feedback is essential for gaining business improvements. BTNI has continued to mature throughout its ongoing self-assessment program. It welcomes the opportunity for external views and areas for improvement within the organization. The company continually seeks to improve the utilization of feedback. The level of feedback integration has been refined from an incrementalist to a strategic approach.

An Organizing Framework

The general organizing framework BI uses for business operations is the Malcolm Baldrige Award criteria. However, as related to customer feedback, the Transactional Customer Satisfaction Index (TCSI) process is the methodology that organizes, integrates, and implements customer feedback. An integrated software system captures the data from all TCSI customer surveys and makes it available for review and analysis. The program links customer requirements to BI's performance. Results of this survey are aggregated and analyzed by BI's business units and market segment, and are communicated monthly to the Business Team, Business Units, and Sales. BI uses a software program to integrate all TCSI feedback and create clear strategic knowledge directed at improving results.

BTNI uses the Northern Ireland Quality Award criteria as its organizing framework. These criteria serve as the method by which to initiate information, which can be integrated and lead to strategic knowledge.

Identification of Strengths and Weaknesses

Although both of the case study companies focus on making improvements to areas of weakness, BI also focuses on the strengths from its customer feedback. BI's customer feedback identifies areas of customer satisfaction as well as areas of customer dissatisfaction. Through use of the TCSI process, BI follows up with customers on recent transactions to determine satisfaction and dissatisfaction. Based on the survey results, action plans are built and implemented with the customer to address any areas for improvement and to continue to improve areas of strength. BI focuses significantly on improving strengths to even higher levels of effectiveness. Again, focusing equally on further improving strengths as well as improving weaknesses is a subtle part of the PDKA logic that can assist companies in moving further into world-class standing.

Clear Decision-Making Criteria

Both BI and BTNI understand the importance of having a clear decision-making criteria to help avoid working on the wrong issues.

BI's TCSI process is reviewed and evaluated regularly by BI's Business Team. The TCSI process itself allows the customer to make decisions that affect the customer, so the decision-making criteria may be different for different customers, but decisions are made, and those decisions are made to benefit the customer.

BTNI uses the Northern Ireland Quality Award criteria to appraise, document, format, assess, and provide feedback to the company. The feedback from the assessment is analyzed and then prioritized. The opportunities for improvement are prioritized into one of the following three groupings: no value, quick fix, or a strategic issue. The opportunities that are of no value are discarded. The opportunities that are determined to be quick fixes are put into an incremental improvement process for those tasks that will take between 1 and 12 months to fix. The opportunities that are targeted as strategic issues are first integrated with other strategic input to formulate an actual strategy. The strategy is then put into the organizational development process, which takes from 12 to 36 months to implement.

Assignment of Tasks

BI uses teams to make improvements. BI has a team-based, high-performance work system supported by an integrated associate development approach. Furthermore, BI's measures from the strategic measurement and analysis model are aggregated and positioned for analysis. Depending upon areas of responsibility, this analysis is performed by a variety of action teams and utilized as the day-to-day management of the business. BI drives change through the TCSI process.

The customized TCSI survey results are reviewed by the account team and the customer. An action plan is created each time a survey is returned. This action plan is built to drive continuous improvement in the services and future deliverables provided to the client. The Delivery Readiness Survey process, which is a part of the TCSI process, culminates in collecting responses and reviewing the responses for concerns that will need to be resolved. When something needs to be resolved, the account manager or travel program manager schedules a meeting to review the results, a team reviews the survey results, creates an action plan, and reviews the action plan and edits as necessary, then schedules and conducts an action plan review with the client and implements based on the client's concurrence.

Periodic Reviews

Both BI and BTNI understand the value of integrating improvement efforts with the business planning and strategic efforts of the company. BI's strategic measurement and analysis model aligns BI's business processes with the strategic plan and the business objectives of revenue growth, productivity improvement, customer satisfaction and Associate satisfaction. BI determines the data and information needed based on its corporate objectives, strategies, core processes, and other key initiatives that result from strategic planning.

BI's strategic measurement and analysis model aligns BI's business processes with the strategic plan and the business objectives of revenue growth, productivity improvement, customer satisfaction, and Associate satisfaction. BI determines the data and information needed based on its corporate objectives, strategies, core processes, and other key initiatives that result from strategic planning.

BTNI has found that assessment feedback should coincide with the business planning process to ensure that results are internalized with normal operations and hence are relevant and timely. To this end, BTNI is actively engaged in developing a strategic model for internalization of the feedback. As BTNI has moved along the continuum of the business excellence spectrum; the sophisticated application and implementation of feedback has prescribed a deepened understanding of the linkages and influences of those feedback reports. The resultant feedback reports have continued to address these fundamentals. BTNI has learned that organizations that adopt a piecemeal approach to strategy formulation and deployment actually suboptimize the overall improvement effort. There is a profound link between the efficiency of feedback integration and the pace and level of improvement.

CONCLUSIONS

Both of the case study companies—BI Performance Services and British Telecom Northern Ireland—have excellent information sources that they use in a comprehensive way that leads to a strategic business knowledge. They achieve this knowledge in slightly different ways, but nonetheless they do achieve it. Each company uses a systematic methodology by which to integrate improvement efforts into its established business planning processes and thus ties its improvement initiatives directly into the heart of its business.

COMPANY

BI Performance Services

PRODUCT/SERVICE:

Performance improvement industry (BI combines communications, training, measurement, and rewards into performance improvement programs for its corporate customers nationwide)

LOCATION:

Minneapolis, Minnesota (headquarters) and 27 offices nationwide

NUMBER OF EMPLOYEES:

1400 associates

KEY PERSONNEL:

Guy Schoenecker: President and Chief Quality Officer
Larry Schoenecker: Executive Vice President
Betsy Schneider: Cast Study Contributor

KEY USES OF INFORMATION:

■ Self-assessment
■ Malcolm Baldrige Award Feedback

ORGANIZING FRAMEWORK TO CREATE KNOWLEDGE:

Malcolm Baldrige Award Criteria

AWARDS AND SIGNIFICANT RESULTS:

■ Minnesota Quality Award, 1994
■ Sears Partners in Progress
■ GTE Directories Supplier Quality Award
■ GM Supplier of the Year
■ Ford Marketing Excellence Award

CASE STUDY:
TRANSACTIONAL CUSTOMER SATISFACTION INDEX

Serving as a strategic partner, BI Performance Services (BI) helps its customers achieve their business goals by improving the performance of people who directly impact those goals: generally a customer's distributors, employees, or consumers. BI is a 48-year-old closely held corporation. Its headquarters are located in Edina, Minnesota, a suburb of Minneapolis. BI's mission is simply **to improve its customers' performance**. BI does this in a variety of ways: improving market share, cost reduction, introducing new products, developing brand loyalty, generating ideas from the employee base to improve productivity, measuring customer satisfaction, recognizing employees, and developing customer loyalty. BI designs and implements these performance improvement programs for its customers using any or all of its four core areas of expertise: Communications, Training, Measurement, and Rewards.

Creativity and innovation are key to providing value to BI's customers. It is how BI attracts and retains customers, and both are customer requirements. Where an approach or service can meet the needs of multiple customers it is referred to as a branded product, but in most cases while the approach may be reused, the specific services are tailored to each customer's specific needs. This custom design approach requires thorough understanding of customer needs, ongoing customer validation throughout the design and delivery process, and careful program management to deliver a quality service on time and within budget.

Scope of Services

Approaches to address issues and improve performance in a customer's distribution channel include:

- Certification programs
- Customer satisfaction research
- New product introductions
- Sales incentive programs, and training programs

Approaches to address employee performance issues include

- Culture or awareness campaigns
- Lead referral programs, quality programs
- Recognition programs
- Safety programs
- Suggestion systems

Consumer issues are addressed through the use of such programs as

- Consumer sweepstakes
- Direct marketing
- Frequency marketing
- Rebates and fulfillment
- Research
- Sales promotion

Other client services that BI provides include

- Worldwide incentive travel programs
- Producing a 160-page merchandise catalog for use in its incentive programs
- Staging large-scale business theatre events
- Designing and implementing training
- Managing sophisticated measurement of its customers' program successes
- Implementing a wide range of research studies
- Providing consulting services to help customers with organizational change, strategic and quality planning

BI SUCCESS STORIES

The following case studies are examples of how BI provides integrated solutions to customers' performance issues.

InfoCenter

The InfoCenter is one of BI's innovative approaches to a customer need. Designed for the automotive market, InfoCenter puts automobiles in populated locations where the public can touch and feel them without the hassle of a salesperson trying to sell something. InfoCenter Advisors, who have been trained on the benefits of the automobile, are available to answer questions. This approach creates a risk-free environment for the consumer where the features and benefits of the product can be explored without sales pressure. In the year that this promotion was rolled out nationally, car sales for the models included in the InfoCenter displays exceeded the corporate sales growth of 2.1% by 5.2%, for a total of 7.3%.

AwardperQs®

An automobile manufacturer needed to increase sales of its compact truck line. There were 2,000 sales managers and 6,000 salespeople to be motivated in the distribution channel. BI recommended AwardperQs, a merchandise, travel, and retail reward system created by BI, to accomplish the goal of increased sales. At the end of the program, the increase in sales exceeded program expectations by more than 15%.

Employee Programs

One of BI's telecommunications customers recently received two awards at the 55th annual Employee Involvement Association Conference for its success with a BI program. This BI customer was recognized as Rookie of the Year for sponsoring the best new employee suggestion program and the Statistical Award for achieving the greatest savings in its industry. To date the program has generated more than 19,000 peer nominations for recognition and more than $40 million in implemented cost savings.

CUSTOMER AND MARKET REQUIREMENTS

BI segments its market based on the audience targeted for performance improvement: distribution, employee, and consumer are its horizontal markets. Further segmentation is by vertical market, focused on an industry rather than an audience. BI's principal vertical markets are Automotive and Telecommunications. Customers are primarily buyers from Fortune 1,000 companies who outsource performance improvement programs.

WHAT MAKES BI UNIQUE

Three major factors are the cornerstones of BI's success: pervasive customer focus, strong Associate focus, and the BI Way. More than 75% of BI's Associates are regularly involved in customer contact. With services custom designed to meet individual customer needs, and long-term relationships with customers, there is intense customer focus throughout the organization. Such approaches as the relationship management and sales processes, the Customer Delight Process for service design and delivery, and the Transactional Customer Satisfaction Index process embody this customer focus. The Customer Delight Process (CDP) is key to BI's success.

The Associate focus reinforces the customer focus. BI Associates are advocates for the company and carry that motivation and pride into their dealings with all customers. Associate productivity is key to the company's financial success. BI has deployed a team-based, high performance work system supported by an integrated Associate development approach. Numerous programs are in place to support Associate well-being, and communication vehicles are used to keep all Associates informed, with compensation, reward, and recognition supporting the objectives and approaches of the company. Mutual respect is widely promoted and practiced. It works: BI customers rave about BI Associates, BI Associates love to work at BI, and company performance validates its effectiveness.

When BI adopted Total Quality Management as a way of life in 1984, the company used the same approach to solving its own business challenge as it would for customers. BI used its performance improvement approach to bring Total Quality Management into the fabric of the company. The BI Way is a process management approach driven by the goal of customer delight. It includes training, problem-solving techniques, process improvement, incentives and a results focus. The BI Way is used by all Associates and is constantly being evaluated and improved.

In 1994, BI became the first service company to earn the Minnesota Quality Award. The Minnesota award has been recognized as one of the most prestigious state awards in the country and is modeled after the national Malcolm Baldrige Award.

In 1998, BI was selected as the only service company nationwide to receive a Malcolm Baldrige site visit, which is the final step of the award process.

BI QUALITY PLEDGE

It is BI's goal to achieve total customer satisfaction through continuous quality improvement . . . satisfaction as defined by each of its customers. Through the achievement of this goal, BI will clearly be identified as the quality leader in its industry.

BI's Key Processes

The strategic measurement and analysis model shown in Figure 1 illustrates the BI business process alignment of the Strategic Plan and the business objectives of revenue growth, productivity improvement, customer satisfaction, and Associate satisfaction. BI determines the data and information needed based on its corporate objectives, strategies, core processes, and other key initiatives that result from Strategic planning. Within the four areas of focus, BI's measures are aggregated and positioned for analysis. Depending upon areas of responsibility, this analysis is performed by a variety of action teams and utilized as the day-to-day management of the business.

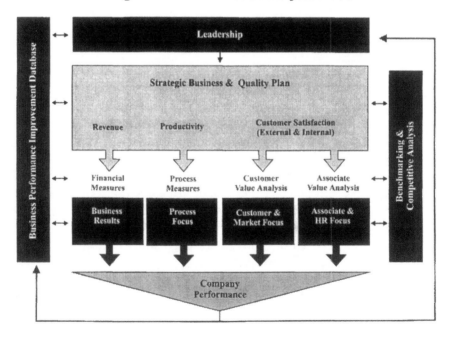

Figure 1   Strategic measurement and analysis model.

Transactional Customer Satisfaction Index (TCSI): An Actionable Survey That is Both Transactional and Customized

The TCSI process is used to determine both customer requirements and satisfaction with key transactions. An integrated software system captures the data from all TCSI customer surveys and makes it available for review and analysis. The program links customer requirements to BI's performance. Results of this survey are aggregated and analyzed by BI's Business Units, strategic account and market segment, and communicated monthly to the Business Team, Business Units and Sales.

The TCSI process integrates two steps to strengthen customer relationships and improve customer satisfaction:

- **Expectations Management**—Asking customers to define their expectations and ways they would like to work with BI prior to program delivery.
- **Transactional Measurement**—Asking customers to evaluate their satisfaction with the delivery of specific events and services immediately following delivery of each of those services or deliverables.

Elements of the TCSI Process include

- TCSI request form
- Customer Expectations Discussion Guide
- Delivery Readiness Survey
- Customized Transactional CSI Survey

The TCSI process is clearly defined through a flowchart, shown in Figure 2, and is a requirement for every event billing $250,000 plus.

BI's TCSI process includes four phases:

- *Delivery Team Meeting (Delivery Readiness Predictor)*—Where BI prepares to talk with the customer. The delivery team assembles to discuss the specifics of what the company is providing to the customer and identifies any open issues via the Delivery Readiness Survey. From here the account team works to develop a Delivery Plan and schedules a discussion with the customer.
- *Customer Expectations Discussion (Expectations Management)*—Where BI clarifies customer expectations about what it is providing to them and how. The Customer Expectations Discussion is the foundation for building a strong working relationship with BI's new customers and the platform for maintaining positive transactions

TCSI Survey Process

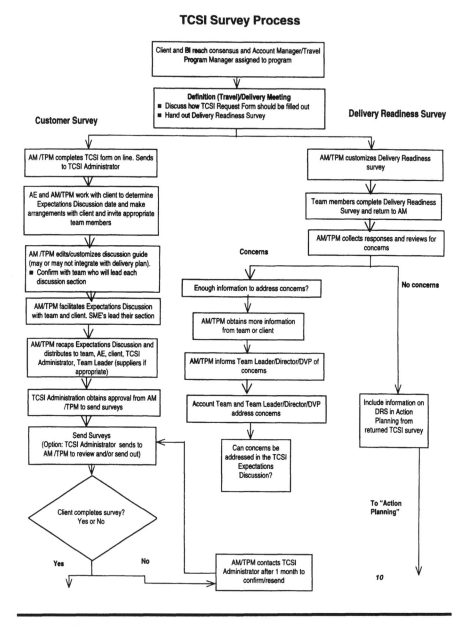

Figure 2   TCI survey process.

with repeat customers. By listening to and managing customer expectations, BI gives assurance that the Account Team wants to do things right the first time and is willing to be measured accordingly.

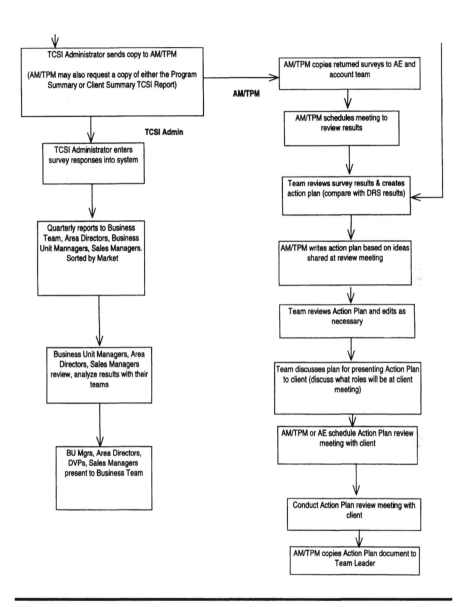

Figure 2 (Continued).

Key members of the delivery team participate in this customer discussion to confirm what services and deliverables BI will provide and clarify specific customer expectations regarding how they wish to work with BI (i.e., Are there any special requirements for approvals on change orders, etc.). At this time BI previews the customer survey. The team asks repeat customers to reconfirm how they like to do business and records any operational or organizational changes that may affect the new program.

The desired outcome is as follows: The Customer Expectations Discussion Recap is completed, recorded, and distributed.

■ ***Survey the Customer (Transactional Measurement)***—After BI completes major phases of delivery, it surveys the customer to gather transactional feedback on how well BI met their expectations. While most programs are surveyed only once, BI has the flexibility to survey customers several times throughout the delivery phase of their project. The TCSI survey is modularized to include general questions asked on every project and questions for each business unit involved in that particular event.

■ ***Using Survey Results (Action Planning)***—From the specific account team working on one project to department and corporate planning, BI uses TCSI results for action planning. Reports are generated based on survey responses received and are distributed right away to account teams and on at least a quarterly basis for department and corporate action planning.

The TCSI is reviewed and evaluated regularly by BI's Business Team. Since its inception, there have been numerous refinements to the TCSI process, including

■ Instituting use of a computer program written specifically for the TCSI to track submitted and maintain a database for results.

■ Developing training sessions on the TCSI process with emphasis on the expectation discussion and action planning. All Account and Program Managers attended advanced expectation discussion training, but in order to be certified they are required to complete the entire TCSI process for at least one project. Advanced Sales Force TCSI training has been completed for the entire field, so that they can readily lead the process.

■ Instituting a pilot to survey account team members regarding their opinions on the team's ability to deliver the program/event before delivery begins via the Delivery Readiness Survey. The purpose of this survey is to identify potential problems and issues before delivery of the program. In addition, this is a potential predictor of future TCSI scores. Based on the pilot outcomes, this survey was enhanced and included in all BI programs.

■ Expanding the TCSI requirements from all events more than $500,000 as a minimum requirement, to include all events more than $250,000 for new clients. Now, the TCSI requirement is for all events more than $250,000.

- Revising Expectations Discussion Guide to make it a more flexible, comprehensive, and useful tool.
- Analyzing the TCSI data since its inception in November of 1995. This analysis led to action planning by each Business Unit at BI. As part of the analysis, an improvement priority scoring system was employed to indicate which specific areas of delivery, when improved, would have the greatest impact on customer satisfaction. The Business Units used this data to build their plans.
- Revising the survey document.

Through use of the TCSI process, BI follows up with customers on recent transactions to determine satisfaction and dissatisfaction. TCSI results are then shared with account team members, including the Account Executive, Account Manager, and other Business Unit Associates as appropriate. Based on the survey results, action plans are built and implemented with the customer to address any areas for improvement and to continue to improve areas of strength.

Since its inception, BI has issued a total of 1019 surveys and received 496 completed surveys for a total response of 49%. The surveys are tracked by BI's Business Units and the results include overall satisfaction by market, compared to American CSI for Service Companies

Figures 3–6 show TCSI results for key drivers by market. BI's results are positive when compared to competitors and the American CSI for service companies.

Through research, BI has determined that the key drivers to customer satisfaction is that it delivers

- On time—by the deadline.
- Within budget—without surprises.
- Accurately—consistent with what was promised.

BI has achieved a satisfaction goal of 8.7 for "accuracy" and "on-time" and an 8.5 for "within budget."

Driving Change through the TCSI

The customized TCSI survey results are reviewed by the account team and the customer. An action plan is created each time a survey (or surveys) is returned. This action plan is built to drive continuous improvement in the services and future deliverables provided to the client. This inevitably leads to a better program overall.

The examples below review the results from customers and the action plan developed specifically from the TCSI feedback.

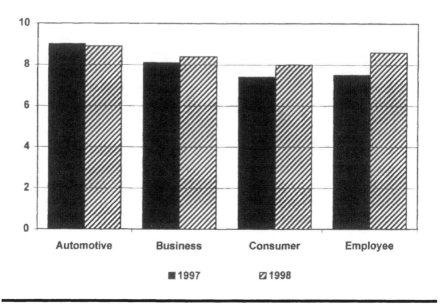

Figure 3   Strategic measurement and analysis model.

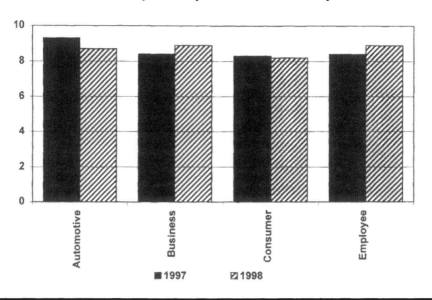

Figure 4   Strategic measurement and analysis model.

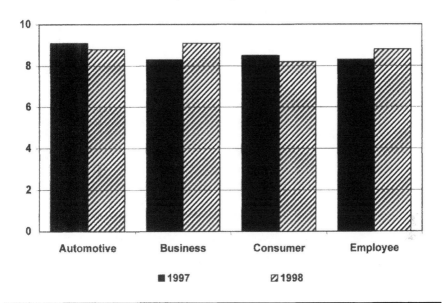

Figure 5   Strategic measurement and analysis model.

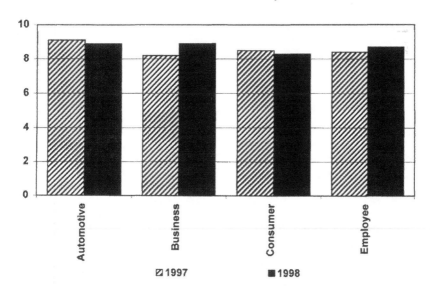

Figure 6   Strategic measurement and analysis model.

Automotive Industry Case Study

BI provided an integrated solution to an automotive company, but changes in the internal and customer team led to some concerns on the delivery of the program. Initial overall TCSI scores were as follows:

Question	Mean Score (scale of 1 to 10 with 10 being most satisfied)
Understood my audience for this project	6.0
Understood my expectations for quality of products and services	6.0
How satisfied were you with BI Performance Services overall	7.0

To address these concerns, the team initiated two quality improvement teams (QITs). The first was to improve the flow of information between internal departments, and included such specific actions as

■ Developing a tracking system for an awards delivery process
■ Designing a system to maintain inventory which could be accessed by all departments
■ Establishing weekly team meetings to communicate proactively on any conflicts or issues which might cause a delay
■ Sharing the responsibility amongst team members in monthly mailing signoffs to ensure accuracy and timeliness of mailings

The second QIT focused on improving the process of delivering awards for current and future programs. A detailed process was developed and agreed upon by team members.

The TCSI results received after the team implemented an action plan showed an increase in all overall categories:

	Initial Mean Score	Scores after Action Plan
Understood my audience for this project	6.0	8.57
Understood my expectations for quality of products and services	6.0	8.29
How satisfied were you with BI Performance Services overall?	7.0	8.53

Health Care Industry Case Study

A health care industry partnered with BI to deliver an integrated incentive program to increase sales. After the event, a TCSI survey was submitted to the client.

The survey results (as outlined in Figure 7) were disappointing. The team developed a detailed action plan for each issue and communicated to its client what action steps it would take to improve its delivery.

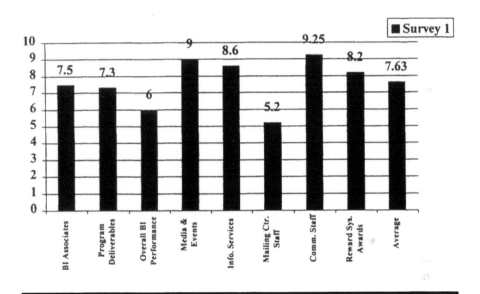

Figure 7 Initial TCSI survey scores.

■ **Issue: BI associates offered helpful options to enhance program.**
Action: Account team members were asked to research all client requests and issues and then provide at least 1–2 recommendations or options.

Score before:	5
Score after:	8.5

■ **Issue: BI program deliverables were delivered within budget, no surprises.**
Action: Account team members are now required to provide written documentation on all new client requests or requirements and obtain

a client signature. Going forward, no verbal commitments/approvals were allowed.

Score before:	3
Score after:	7.5

■ **Issue: Overall BI had not enough concentration on processes.**
Action: The account team developed process flowcharts of internal process for the account and provided this information to the customer.

Score before:	6
Score after:	8.5

The survey results after the Action Plan are compared with the original survey results in Figure 8.

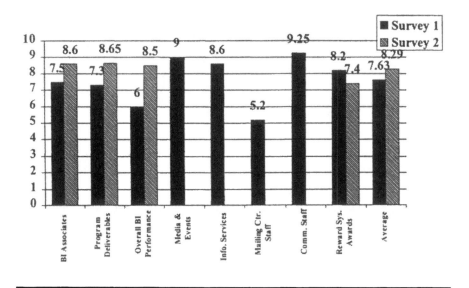

Figure 8   TCSI scores after Action Plan.

Next Steps

BI's latest TCSI overall score was at 8.6 on a 10-point scale. This score compares well to scores for BI's competitors and other service companies, but BI continues to refine its TCSI process in key ways, such as

- Analyzing and reviewing the current survey vehicle, looking for ways to make it easier and more convenient for customers to complete
- Determining areas of the survey which are key factors in customer satisfaction
- Providing additional tools to BI associates to promote the advantages and outcomes of this process to BI's customers
- Providing essential information to BI's Business Team and Account Team members to help drive the process and improve results

COMPANY

British Telecom
Northern Ireland

PRODUCT/SERVICE:
Telecommunications

LOCATION:
Belfast, Northern Ireland, United Kingdom

NUMBER OF EMPLOYEES:
2,500

KEY PERSONNEL:
Russell Simpson, Edel O'Neill, Sarah Meegan, Norman McLarnon

KEY USES OF INFORMATION:
- ISO 9001
- Self-assessment

- British Telecom Award Feedback
- Northern Ireland Quality Award Feedback
- United Kingdom Quality Award Feedback

ORGANIZING FRAMEWORK TO CREATE KNOWLEDGE:

Northern Ireland Quality Award criteria

AWARDS AND SIGNIFICANT RESULTS:

- ISO 9001 Certification
- Northern Ireland Quality Award
- United Kingdom Quality Award

INTRODUCTION AND BUSINESS OVERVIEW

Embracing the Feedback: An Award Winner's Experience

In recent years, self-assessment has been the subject of many research projects and surveys. However, few of these studies have examined the reaction or receptiveness of quality award applicants to external feedback reports from the award assessors. It is the contention of this study that such reactions have a significant impact upon the integration of assessment feedback into company-wide operations and activities. Case study analysis of British Telecom Northern Ireland (BTNI) is presented to provide some insight into the experience of a prominent quality organization in this context.

British Telecom Northern Ireland (BTNI)

BTNI entered the 1994 U.K. Quality Award on the crest of a wave. This former monopoly had successfully come to terms with competition and, in 1993, had been the first operational unit of BT to achieve ISO 9001 registration. To continue the improvement process, BTNI had adopted self-assessment in 1993, winning both the BT Group and Northern Ireland Quality Awards within 12 months. Having entered the U.K. Quality Award with realistic expectations of winning, BTNI traversed the first three phases of the transition model (shock, denial, depression) rapidly, if painfully, and on reaching phase 4 (letting go, acceptance of reality) decided to re-enter the 1996 U.K. Quality Award.

With notable organizational improvements in place and a more professional approach to the award process, the disappointment felt at not winning was, if anything, even more acute than in 1994 and, briefly, the organization regressed to denial. However, it is a tribute to the professionalism of the BTNI management that these feelings soon dissipated and, buoyed up by winning 1996 Northern Ireland Quality Award, the organization moved to phases 5 (testing) and 6 (consolidation). That the approaches adopted were successful was recognized when BTNI was a winner in the 1997 U.K. Quality Awards.

Organizational Response to Successive Feedback Reports

It all seems so logical. Take a structured total quality model (Malcolm Baldrige or European Business Excellence), prepare a balanced presentation of the organization, and submit it to the award-granting body. Meanwhile, the awarding body selects and trains experienced individuals who are capable of assessing organizations. After much time spent by the assessment team poring over the submission, debating its merits and reaching consensus, the lead assessor will prepare an objective feedback report clearly identifying the applicant's strengths and areas for improvement (AFI). It is generally assumed that the applicant will use the feedback to identify the systematic factors and process weaknesses that are responsible for poor performance. However, BTNI contends that the initial response of the applicant's senior management team to the feedback report is crucial for the future success of self-assessment within the organization.

That modern managers are explicitly committed to the success of their organizations and have a heavy personal involvement is not in doubt. However, personal commitment can lead to emotional reactions to external feedback and may present a barrier to the ongoing success of self-assessment and involvement. The managerial team of an effective and competitive organization, who genuinely believe they are potential award winners, will be surprised to receive a lengthy list of areas for improvement (AFIs) and, consequently, what they perceive to be a low score: less than 500 points out of a possible 1000 points. To overcome this barrier, BTNI proposes that management employ a transition phase model, based on the work of Adams et al. (1976), to assist them in coming to terms with the initial feedback report and further, in progressing their program of assessment. Drawing upon the transition phase model, organizational reactions to external feedback are conceptualized as follows.

- *Phase 1: Shock* ("This cannot be my organization"). No matter how tactful, diplomatic, and balanced the feedback report is, an objective analytical list of the AFIs in the organization will come as a shock. This is reinforced by the very human reaction of most recipients, who will ignore documented strengths and focus entirely on the score and AFIs. A more fundamental issue is that, upon entering the award process, most organizations believe that assessment will be based on what they are actually doing in practice, rather than against the criteria of very demanding models such as the Malcolm Baldrige or European Business Excellence models. General management tends to underestimate the difficulty of realizing a high score. In general the initial impact of the self-assessment report is likely to undermine management's ability to take in the information and act constructively. A normal reaction is to lay the feedback aside and lose valuable time.

■ *Phase 2: Denial* ("This is not my organization"). Ignoring the proven effectiveness of self-assessment as a process and the many hours the external assessment team will have devoted to analyzing the applicant, this phase is typified by a retreat from the need for organizational change. Resistance to change will be highest during this period. This is a particularly dangerous phase; unless these feelings can be overcome, self-assessment as a diagnostic tool will encounter little success in the organization, even if insisted upon by corporate headquarters as organizational policy.

■ *Phase 3: Disillusionment* ("This possibly is my organization"). While the reality of the need for organizational change becomes apparent, the ability of self-assessment to identify widespread "weaknesses" throughout; an organization may result in management believing that effective corrective action is outside its control. This phase can be a disillusioning period, and could result in management abandoning self-assessment altogether and searching for an organizational improvement tool that may be easier to comprehend.

■ *Phase 4: Letting go* ("I have got to improve my organization"). This phase involves accepting the need for organizational change and recognizing that the information contained in the feedback report can play a vital role in the organization's improvement process.

■ *Phase 5: Testing* ("What can I do to improve my organization?"). This phase reflects an active and creative response to the self-assessment feedback. More energy and enthusiasm are available, but self-assessment may be blamed if measurable improvements are not apparent within a short time span. A normal approach is to select and allocate "raw" AFIs to members of the management team for correction. In most organization, the improvement process is unlikely to be systematic at this stage.

■ *Phase 6: Consolidation* ("I can manage improvement better"). Out of the testing process come methods of utilizing the feedback which become accepted as norms. Reflecting on the experience gained so far, it is recognized that there is greater merit in utilizing the self-assessment report than merely allocating AFIs. Possible responses are to screen the "raw" feedback report, identifying and separating strategic issues, while, in parallel, systematically implementing short-term incremental improvements.

■ *Phase 7: Internalization* ("Self-assessment is a process which we use effectively and routinely for organizational improvement."). During this mature phase of development, the use of the feedback report, as a diagnostic tool, is routine and unthinking. It is at this point that the learning and growth which self-assessment provides are recognized and built on.

BTNI's Experience of the Transition Phase Model

Conceptualized as a gauge of organizational response to external feedback, the transition phase model outlines the various stages through which the management team will progress during successive self-assessment exercises. The success of BTNI in this context is a measure of how quickly and effectively the management team was able to work through the first three phases of the model. This, in turn, demonstrates the commitment of the company to a philosophy of continuous improvement.

In general, the level of senior management commitment to the process of self-assessment is manifest in the early stages of the transition model. Failure to accept the external feedback reflects misgivings in the rigor of the entire self-assessment process and follow-up exercises are unlikely to be undertaken. Those companies that do progress, on the other hand, have learned much from their first exercise. In the case of BTNI, the company integrated its second feedback report from the standpoint of phase 4, "letting go." This reflected an acknowledgment of the robustness of the complete self-assessment cycle and the need to act on the feedback in order to advance.

BTNI has continued to mature throughout its ongoing self-assessment program. It welcomes the opportunity for external views and areas for improvement within the organization. The company has reached the consolidation phase of the model and continually seeks to improve the utilization of feedback. The level of feedback integration has been refined from an incrementalist to a strategic approach.

Integration of Assessment Feedback

In recognition that a static approach to feedback integration results in decelerated rates of improvement, the propensity for improvement from ongoing self-assessment exercises is limited to the momentum in place. This is supported by Van der Weile et al. in the state-of-the-art study on self-assessment published in *The TQM Magazine* (1996, pp. 13-17), who noted of organizations which had undertaken more than one self-assessment exercise: "Only minor changes have taken place in the procedures, people/unit involved in the assessment process, the scoring criteria, and the communication/feedback issue."

As BTNI has moved along the continuum of the business excellence spectrum, the sophisticated application of the model has prescribed a deepened understanding of the linkages and influences. The resultant feedback reports have continued to address these fundamentals. Organizations adopt a piecemeal approach to strategy formulation and deployment suboptimize the overall effort. There is a profound link between the efficiency of feedback integration and the pace and level of improvement.

Implicit in the final stage of the transition phase model is that the scheduling of the assessment should coincide with the business planning process to ensure that results are internalized with normal operations and, hence, relevant and timely. To this end, BTNI is actively engaged in developing a strategic model for internalization of the feedback. Figure 1 demonstrates how BTNI envisions the self-assessment cycle feeding into two key areas. The organizational development cycle represents planning on an extended time-frame, and recognizes the strategic implications of the report. Throughout this process, the feedback is internalized and subsequent strategies are deployed effectively through Kaplan and Norton's (1992) balanced business scoreboard. The incremental improvement cycle circumvents this process, allowing an expedient approach to one of improvement activities.

CONCLUSIONS

It is widely recognized that subsequent self-assessments are more successful than the first exercise. For many organizations, this success stems from a combination of improved evidence-gathering and the introduction of new approaches and measures that hitherto were not recognized as critical to the business. This strategy is favored by risk-averse companies that seek to depersonalize the outcomes of the initial assessment, and is characteristic of organizations in the early phases of the transition phase model reconceptualized in this context. The study has demonstrated that

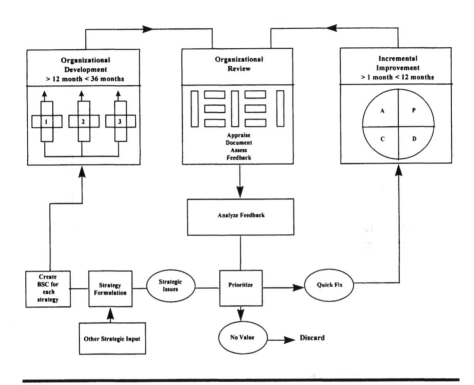

Figure 1   Managing organizational strategic development.

BTNI has taken a pragmatic approach to the management of this feedback and hence has progressed effectively to the latter phases of this model.

In general, as an organization acquires a higher degree of proficiency in business excellence, the rate of improvement and the associated trends in assessment scoring decelerates. Sustaining the momentum at this stage of the self-assessment journey mandates that the feedback integration is coordinated, internalized, and strategically deployed. As BTNI has recognized, at this stage the organization will have reached the final phase of the transition phase model.

9

HOW TWO WORLD-CLASS SERVICE ORGANIZATIONS IMPLEMENT PRINCIPLES OF PDKA— ROYAL MAIL AND THE UNITED WAY OF MIDDLE TENNESSEE

INTRODUCTION

The companies that provided these case studies—Royal Mail and the United Way of Middle Tennessee—differ in the services they provide. Royal Mail is a public (government of the United Kingdom) organization. The United Way of Middle Tennessee is a not-for-profit organization that has implemented a significant portion of PDKA logic into its business systems.

Royal Mail

Royal Mail's services are the collection, sorting, and delivery of first- and second-class mail, as well as priority services, electronic services, special collection and delivery services, direct mail services, response services, and international services. Royal Mail makes a crucial contribution to the British way of life, ensuring that even the most remote parts of the country are not isolated and helping keep communities alive. A special character-istic of Royal Mail's business is that everyone either sending or receiving a letter is a customer. Hence, Royal Mail is part of a wider network of communication in society.

Royal Mail is winner of the European Quality Award, and is now being used as a benchmark organization for certain processes. The ways in which the organization created and used its Leadership Charter to raise business performance is just one accomplishment other companies seek to emulate. Overall, there have been many lessons learned, two of which line up with the PDKA logic: the value of benchmarking with organizations in different sectors on their performance, processes, tools and techniques used (gaining knowledge), and the need to exact the maximum amount of information from the vast amounts of quality-related data collected from all over the organization, regarding performance and perceptions.

Strategy at Royal Mail is driven by the vision and business direction and goals, with their associated results, targets, and measurements. There are three management processes for formulating strategy and delivering the results. The strategic direction setting process occurs first, and is carried out by the Royal Mail Executive Committee. Targets are developed via the national and business unit planning processes. The planning process synthesizes marketing plans, national planning guidelines, and local requirements, turning them into consistent and achievable plans. The strategic business units turn them into achievable plans. The results are then reviewed systematically.

The United Way of Middle Tennessee

The United Way of Middle Tennessee's (UWMT) function is to raise and distribute resources to local non-profit health and human service programs. The mission of United Way of Middle Tennessee is to be the single most efficient way to identify, address, and support solutions to local community health and human service needs. A key ingredient in UWMT's success is its use of volunteers. More than 2,500 volunteers provide direction and resources to help the United Way accomplish the mission and strategies.

UWMT has 2 years of results that reflect the implementation of the Community Solutions Council process and the integration of assessment feedback with other organizational feedback tools. Since the implemen-tation of its strategic plan and the integration of business planning with its feedback implementation process, it has exceeded overall campaign growth each year. UWMT has also won the Excellence in Service Quality Award (ESQA) Bronze Level Award and the Tennessee Quality Award—Commitment Level.

HOW THESE COMPANIES HAVE IMPLEMENTED THE PRINCIPLES OF PDKA

Executive Buy-In

One of the challenges originally facing the United Way of Middle Tennessee was to convince senior managers that the behaviors it models were critical to the success of the performance excellence journey. In short, a radical cultural change had to occur that would be led by the President and the senior managers. UWMT relied on Dr. W. Edwards Deming's teachings as the foundation for this new approach to managing its business. A key point for which UWMT should be proud is that its action plan team, which pulls together all of the organization's information, is composed of none other than its senior leaders. The senior leaders actually reviewed the feedback reports from UWMT's Tennessee Quality Award feedback and Excellence in Service Quality Award feedback. So, UWMT's senior leaders were put into an active position to clearly understand the information, integrate the information, and make decisions about the new knowledge.

Action Plan Team

UWMT's senior leaders make up the action plan team. This action plan team pulls all of its organizational data together, and does it in concert with UWMT's business planning process. UWMT's senior leaders review the organizational data sources, such as donor surveys, customer feedback collected, employee surveys, strategic plan progress, and agency surveys.

Consolidation of Information

Royal Mail uses information for a business advantage. Policy and strategy cannot be effectively determined without information derived from carefully collected feedback. Royal Mail uses four main sources of feedback.

1. Feedback from customers, expressed as customer perception and satisfaction index data and details of how customers are handled at Customer Service Centers are used to improve performance at a national level.
2. Suppliers are also listened to carefully.
3. Feedback from employees, via the opinion survey, is used nationally at all levels.
4. Data on social, regulatory, and legislative issues, and economic indicators are reviewed.

Key measures are examined at Royal Mail: employee opinion surveys and the Customer Satisfaction Index, as well as an end-to-end service measurement system. Underpinning this system is a process to obtain feedback on managerial behavior that would naturally affect service performance. The benchmarking technique of examining best practice from other organizations has been applied to the business excellence processes. The view now from the top of the organization is that the process of benchmarking has been "the most effective catalyst for change."

UWMT's key uses of information include quality award feedback (the Tennessee Quality Award and Excellence in Service Quality Award), donor face-to-face feedback (a customer satisfaction index) volunteer and staff feedback, and Critical Organization Indicators. UWMT's management team consolidates the feedback information from its quality award feedback and other organizational sources, then it groups the information into opportunity areas. The decision was made to use a modified version of the feedback integration process. The timing of this review process was coordinated with the management team's creation of the annual business plan, which supports the organization's strategic plan.

Organizing Framework

Royal Mail chose the EQA model because it offered a framework consistent with what was already being done internally. The organization is currently focusing on three areas: leadership, business processes, and self-assessment.

UWMT's organizing framework to create knowledge is its Excellence in Service Quality Award criteria using the Plan-Do-Knowledge-Act model. A matrix is used by the management team to consolidate feedback into opportunity areas. All feedback items regardless of source are arranged under the appropriate Excellence in Service Quality Award (based on the Baldrige) categories.

Identification of Strengths and Weaknesses

Business Excellence Reviews, self-assessments based on the European Quality Award model, are a widely used tool at Royal Mail. Together and individually, these assessments highlight strengths and improvement opportunities for the relevant units.

Clear Decision-Making Criteria

UWMT's management team makes the decisions on what to proceed with and prioritizes the opportunity areas.

Prioritization of Strengths and Weaknesses

UWMT's management team uses a matrix to consolidate feedback into opportunity areas and prioritize the opportunities. All feedback items, regardless of source, are arranged under the appropriate Baldrige categories. The matrix lists the types of information in columns so that the management team can check off the similarities and differences among information sources. The matrix is completed by each management team member. Scores for each item are totaled and averaged for the team. The items are then ranked from high to low. A discussion of each item takes place with consideration given to priority from the score, the organizational capability to implement the improvement opportunity, and additional human or financial resources required for implementation. The team decides which items to work on for the coming calendar year. Typically, any item with a score of 4 or higher on the priority list is worked on in the coming year. The relatively small staff makes it easier for the management team to quickly assess resource capabilities and action steps necessary for improvement strategies.

Periodic Reviews

After the opportunity areas have been prioritized, UWMT's management team produces the business plan and the budget for the organization. Since the same management team has personally reviewed all of the organizational information and determined opportunities for improvement, those improvement opportunities are factored into the business plan, and budgets are allocated for the high-priority improvements. Once the plan and budget are developed, the volunteer strategic planning committee and Board review and approve the business plan and budget. When the plan and the budget are approved, department and individual workplans are developed to support the business plan. And finally, when the plan is implemented, quarterly reviews of progress by the management team and the Board take place. The ongoing collection and analysis of customer feedback completes the feedback loop. UWMT applies for at least one Baldrige-based award each year in order to receive the feedback reports, which are also integral to the feedback integration process cycle.

CONCLUSION

Royal Mail has recorded a continued rise in the return on capital. The amount of profit per employee has risen due to automation and labor productivity gains, while the overall number of employees has declined, despite the increasing volume of items being handled. Royal Mail has achieved 12 successive years of volume growth of national mail, despite the economic recession in the U.K. and the rapid growth of other forms of communication, such as fax and electronic mail.

UWMT is a unique case study, partly because it is a not-for-profit organization, which means that it does not have a product or service it sells for revenue generation like most other businesses. It relies on donations and it is trying to make these donations stretch further and further by improving efficiencies of its business operations and key processes. UWMT has used the PDKA logic, modifying it slightly to match its organizational realities and needs. The process of integrating feedback from various sources and integrating this information into the annual business and strategic planning process has enabled UWMT to focus on priority issues within the framework of the strategic plan.

COMPANY

Royal Mail

PRODUCT/SERVICE:

Collection and delivery of first- and second-class mail, as well as priority services, electronic services, special collection and delivery services, direct mail services, response services, and international services

LOCATION:

United Kingdom

NUMBER OF EMPLOYEES:

160,000

KEY PERSONNEL:

David Peden, Geoff Carter of the European Foundation for Quality Management, and the European Commission

KEY USES OF INFORMATION:

- Employee opinion surveys; feedback from suppliers; customer satisfaction index; data on social, regulatory and legislative issues, economic indicators

- Business Excellence Reviews (self-assessment)
- Benchmarking
- "End to end" service measurement system
- Adoption of the European Quality Award Model as the organizing framework

ORGANIZING FRAMEWORK TO CREATE KNOWLEDGE:

European Quality Award Model

AWARDS AND SIGNIFICANT RESULTS:

- Rise in return on capital employed 2.2% to 20% (5 years)
- Reduction in real unit costs of 2.8% (5 years)
- Profit per employee has increased approximately 950% (5 years) to about 1,900 pounds per employee
- 12 Successive years of volume growth of national mail
- Winner of European Quality Award

INTRODUCTION AND BUSINESS OVERVIEW

Royal Mail is more than just a business. It makes a crucial contribution to the British way of life, ensuring that even the most remote parts of the country are not isolated and helping to keep communities alive.

A special characteristic of the business is that everyone is likely to be a customer, either sending or receiving a letter. Hence, Royal Mail is part of a wider network of communication in society.

Responsible for the collection, sorting and delivery of letters and packets, Royal Mail delivers over 16 billion letters a year and over 64 million per day. Mail is collected from up to 120,000 points all over the country every day, and delivered to over 25 million U.K. addresses, 6 days a week.

This service is operated by some 160,000 employees working from over 1,900 sites, and, with a fleet of 27,500 vehicles, travelling 400 million miles annually. They generate a turnover in excess of £4 billion with a pretax profit of over £300 million.

How Royal Mail is Organized

Royal Mail is part of the U.K. Post Office Corporation, whose sole shareholder is the government. The four principal businesses are:

- Royal Mail (letter and packet business)
- Parcel Force (parcel distribution)

- Post Office Counters (a retail network of post offices with around 20,000 outlets delivering mainly financial services to the public)
- Subscription Services (a national organization providing licensing and subscription management services)

Royal Mail itself comprises 17 business units (9 divisions, 4 strategic business units, and 4 business centers) and Royal Mail Strategic Headquarters.

Overall leadership and direction for Royal Mail is provided by Strategic Headquarters, which also ensures that the interfaces with other businesses within the Post Office Corporation are managed effectively.

Three of the strategic business units are involved in integrated market and product development: national (for domestic U.K.), international (world-wide), and streamline (for organizations posting large volumes of mail). Cashco is the fourth, and operates in the cash security market.

The divisions are large organizations in their own right, with as many as 20,000 employees, and annual turnovers of up to £700 million. They deliver the Royal Mail products and services to the customers.

The business centers offer support services to Royal Mail and to other business units within the Post Office. For example, Quadrant provides catering for the whole of the Post Office Group and RoMEC does design consultancy, engineering, manufacturing, and provides security cover.

The Consultancy Services Group offers a range of central support services, such as project management, information systems development, market research, and other specialist expertise, mainly for Royal Mail.

Characteristics of the Market

Royal Mail operates in the communications market, transferring information, funds and personal messages nationally and internationally. Its mission is "to be recognized as the best organization in the world distributing text and packages." The entire U.K. market is estimated at £30 billion, of which Royal Mail's share is around 17%.

INCREASING COMPETITION

Although part of the business is protected by a statutory letters monopoly, applicable to all letters charged at under £1, competition is fierce. The majority of the business is open to the competitive forces of the marketplace. The established forms of information transmission, such as telephone, fax, television, and advertising have now been joined in the marketplace by various forms of courier services, electronic data interchange, and electronic mail. Possible further deregulation at home and increasing competition from abroad will demand greater efforts of Royal Mail to maintain and grow its market share.

PRODUCTS AND SERVICES

The service most frequently associated with Royal Mail is the collection and delivery of First and Second Class Mail, arriving the next day and 3 days later, respectively. There are also six other types of service, designed to meet the needs of individuals and businesses:

- Priority Services (special, recorded, or registered delivery)
- Electronic Services (electronic post, fax, mail, and electronic data interchange)
- Special collection and delivery services (collection direct from a business or redirection of mail)
- Direct mail services (designed for businesses that post large numbers of letters or packets and specially prepare mail for discounts)
- Response Services (business reply and Freepost for customers to respond to promotional campaigns)
- International services (for posting letters, packets and other items overseas under such means as Airmail, Swiftair, and Surface Mail)

IMPERATIVES FOR CHANGE

The Pursuit of Excellence

Back in 1988, then as now, Royal Mail was making a substantial profit. However, internally there were misgivings. Market share was being eroded, customers were not satisfied with the service they were receiving, and morale was not high.

In 1988, senior management held a series of "away days" to focus their priorities and developed clear statements on the current and future desired state of Royal Mail. At this stage, ideas were just being formulated, so the group interviews (conducted by the Quality Director) were restricted to the top 150 managers, with approximately 10 people in each focus group. From these discussions emerged six categories of issues on which the organization would work and which formed the platform for the quality initiative.

Mission and Values

As a part of this organizational self-analysis, the group acknowledged that there was no explicit mission statement to answer the question, "Why does Royal Mail exist?" Therefore, any definitions of quality could only occur in a vacuum. The Board of Directors addressed this issue and produced a business mission: "As Royal Mail, our mission is to be recognized as the best organization in the world distributing text and packages." We shall achieve this by:

- excelling in our collection, processing, distribution and delivery arrangements
- establishing a partnership with our customers to understand, agree and meet their changing requirements
- operating profitably with efficient services which our customers consider to be value for money
- creating a work environment which recognizes and rewards the commitment of all employees to customer satisfaction
- being forward looking and innovative

Significant Milestones

It is impossible to detail here all the efforts to improve business performance and change the culture. However, a number of critical points marked the successful path to excellence. These seven major initiatives are set out in the following paragraphs.

First, key measures had to be examined. Employee opinion surveys and a Customer Satisfaction Index were introduced, as well as an "end to end" service measurement system. Underpinning this system was a process to obtain feedback on managerial behavior that would naturally impact on service performance.

Next, a suitable infrastructure had to be developed. Between 1989 and 1990, Royal Mail recruited 85 quality support managers. These reported to key senior managers to advise on quality-related issues.

Around the same time, the organization began a program to create a continuous improvement culture. First, all senior and middle managers held 5-day workshops with their teams, cascading down the organization from the managing director's own team. These workshops, based on Customer First concepts and with a required output of an improvement project, led to hundreds of quality-driven activities. The current success of the change in culture is demonstrated in the annual Teamwork Event, attended by over 15,000 employees, customers, and suppliers, which displays the best of the improvement activities across the organization.

A fourth key milestone was the benchmarking of the Total Quality Process itself. Groups of senior managers visited organizations in the United States that had won the Malcolm Baldrige National Quality Award or were close to doing so. These visits confirmed that Royal Mail was heading in the right direction and offered some new insight into changes in organization and processes needed to enhance a quality service.

Fifth, management felt front-line involvement was crucial; from 1992, all 120,000 employees attended Customer First Awareness sessions on the meaning of customer-focused performance improvement.

Also in 1992, a drastic simplification of the organizational structure reduced the number of business units from nearly 70 to 19 and the size of the headquarters by 90%. The result of all this was to devolve more and more responsibility to the core business.

The seventh major milestone was the adoption of the European Quality Award model in 1992. Royal Mail chose the EQA model because it offered a framework consistent with what was already being done internally. The organization is currently focusing on three areas:

- leadership
- business processes
- self-assessment

Clearly, for an organization of this size and geographical dispersal, the move to adopt a customer-oriented culture involved a massive investment of time, energy, and resources. The sheer logistics of ensuring that the quality message was heard by 160,000 employees, followed by a process of assimilation and commitment was, and still is, a challenge to the organization.

PROCESSES

Strategic Direction and Business Development: Formulating Strategy

Strategy at Royal Mail is driven by the vision and business direction and goals, with their associated results, targets, and measurements. An interesting illustration here is how part of the mission (caring about their role in the community and their impact on society) feeds into the vision ("to be an active member of society"). This, in turn, influences the directional goal of "being a responsible corporate neighbor." Consistent with all this is the marketing concept that one of the brand values is "to be genuinely caring" and one of the brand characters is "to build fulfilling relations." These concepts are important to Royal Mail as a commercial basis on which to operate. The company wants its customers to see it as a safe and secure service to which they can safely entrust things of importance and value to them. If Royal Mail is seen to have employees who care about the community at a local level, then "it helps to support the kind of corporate image Royal Mail wishes to cultivate. We also end up with a more complete person—one who is of value to us" (Assistant Managing Director).

There are three management processes for formulating strategy and delivering the results. First, the strategic direction setting process is carried out by the Royal Mail Executive Committee. Targets are developed via the national and business unit planning processes. The planning process synthesizes marketing plans, national planning guidelines, and local requirements, turning them into consistent and achievable plans. The strategic business units turn them into achievable plans. The results are then reviewed systematically.

The Royal Mail Executive Committee also leads cross-functional forums whose members are predominantly from the business units. They own elements of the EQA model and ensure relevant data is fed into policy and strategy and its implementation. These forums provide the basis on which the three management processes operate.

USING INFORMATION FOR BUSINESS ADVANTAGE

Policy and strategy cannot be effectively determined without information derived from carefully collected feedback. Royal Mail uses four main sources of feedback.

Feedback from customers, expressed as customer perception and satisfaction index data, and details of how customers are handled at Customer Service Centers are used to improve performance at a national and local level. Each Executive Committee director is matched to at least one major account customer establishing a partnership at a strategic level.

Suppliers are also listened to carefully. For example, AEG, a manufacturer of automated mail processing equipment, was consulted to develop related technological capabilities and market opportunities.

Feedback from employees, via the opinion survey, is used nationally at all levels. For example, when the national employee strategy steering group realized that employees were dissatisfied with the recognition they received, the group commissioned a study of recognition practice within business units. One division has front-line feedback teams reporting to the General Manager.

Data on social, regulatory, and legislative issues, and economic indicators are reviewed at headquarters. An annual environmental scan highlighting topical issues is made available to senior managers and forums undertaking strategy formulation and planning. For example, a recent national survey, conducted by MORI, identified four different areas of concern to the general public: emissions (e.g., carbon dioxide), vehicle pollution, waste disposal, and the use of natural resources. Royal Mail has responded to this with its own studies and has achieved significant results in certain areas.

QUALITY INFORMATION FOR A QUALITY BUSINESS

Royal Mail uses information technology to support business processes. A cross-functional steering group at executive director level has developed a strategic approach to ensure that information technology is exploited for competitive advantage. This approach also draws on information technology strategy from the Post Office.

Information is captured only once, and is available in an accurate, timely, and consistent form for all who need it. Information is seen as an organizational resource, and systems are designed to be easy to use. The involvement of the various business units has resulted in several process improvements, including the development of a "professionalism" policy for information systems people, and a computer literacy program for managers. The next stage of organizational learning is to assess how effectively the various managers and working groups are using information systems to improve the quality of service.

Computing and information processing is carried out within national computing security guidelines. Ensuring conformance with these guidelines is the responsibility of the business unit information systems managers. At the same time, Royal Mail recognizes that to be of use, information must be accessible.

Customer processes have been a very high priority since 1994, and Royal Mail has developed many application systems in this area. These include a customer care computer system and a mail redirection computer system. Other applications are being developed to integrate processes into customer and supplier environments.

Business Excellence Reviews

Self-assessment, based on the EQA model, is a widely used tool at Royal Mail. Together and individually, these assessment exercises highlight strengths and improvement opportunities for the relevant units. They are carried out confidentially by internal staff who are trained assessors. The results are used only by the unit organization under review. The top 200 managers have all been trained as assessors. One assistant managing director states that

It is the best form of management training ever given because it relates to the complete model of a business and how it works. It trains people to critically evaluate a business and look for the interactions. It has also provided a cadre of people who understand it and who are committed to deploying it. It is regarded as a way of working.

The entire review process is regularly updated and improved. The current policy is for each unit to be assessed every 2 years, sharing good practice and establishing a close link between the assessment results and business planning. Desired improvements form part of the next planning round.

The process has now been modified and extended to individual units (called Unit Excellence Reviews) and offices to establish it as a normal part of working routine. At this level, its purpose is to encourage involvement in improvement activities, assess progress, and recognize excellence and achievement. The organization has learned that the review process also enables the sharing of good practice and improves teamwork and empowerment. Self-assessments started relatively recently, and Royal Mail is aware that these will have to settle into the normal work routine.

Benchmarking

This technique of examining best practice from other organizations has been applied to the business excellence processes, as well as other processes. In the former case, TSB, British Telecom, ICL, and Kodak were partners. In the early days of total quality at Royal Mail, some six trips over 3 1/2 years were made to the U.S. to visit Malcolm Baldrige National Quality Award-winning companies such as Milliken, Westinghouse, Motorola, and Xerox. The view now from the top of the organization is that the process of benchmarking has been "the most effective catalyst for change."

Results: Financial

Royal Mail has recorded a continued rise in the return on capital employed from 2.2% in 1989 to 20% in 1993–4. This measure reflects the successful management of profit, working capital, and utilization of fixed assets. In these years the government target has been met or exceeded. There has also been a reduction in real unit costs of 2.8% in the 5 years to 1993–4, although significant sums have been invested in laying the foundations for a strong quality culture.

The amount of profit per employee has risen, due to automation and labor productivity gains, while the overall number of employees has declined, despite the increasing volume of items being handled.

The results are impressive, in that Royal Mail has had 8 consecutive years of profit, but the organization recognizes that it will need to compare itself with others in the industry sector to provide a truly accurate benchmark of performance.

Results: Nonfinancial Measures

Nonfinancial measures of business success are also very important in their own right, as well as underpinning the financial results. A major element of customer satisfaction is the reliability or quality of service provided. Royal Mail agrees on the target for first class letters with a government-appointed consumer advocate group called the Post Office Users National Council. Quality is the speed and reliability of letter mail, using end to end sampling (time taken from posting anywhere in the country to receipt by customer) carried out by an independent operator. National targets are then devolved to Royal Mail's nine divisions. The service level for every product is then tested using 34,000 samples per month.

Great improvements have been made since 1989 when this check first began. Then, approximately 78% met this target. In 1993, 91.2% of first class letters were delivered on the first working day after posting. The target continues to increase.

The improvement in first class letter performance is mirrored in Royal Mail's other national products. The comparison with other major European post offices, carried out by an independent market research company, Research International, is also noteworthy.

Royal Mail has achieved 12 successive years of volume growth of national mail, despite the economic recession in the U.K. and the rapid growth of other forms of communication, such as fax and electronic mail. However, there has been a very slight decline in the overall share of the total communications market, where competition is extremely fierce.

Conclusion

In conclusion, it may be said that even with all this enormous and sustained effort, Royal Mail believes that there is still more it needs to do to become a total quality organization. The sheer logistics of attempting to do this in an organization of this size are daunting. It is not surprising that, initially, the speed at which ideas and concepts can be fully deployed varies and some may take up to 5 years to fully deploy.

There are many lessons that have been learned and many that are being shared with others not as far along the road to quality. Royal Mail is now being used as a benchmark organization for certain processes. The ways in which the organization created and used its Leadership Charter to raise business performance is just one activity other companies seek to emulate.

Lessons Learned

Overall, the lessons that have been learned are

- The importance of continually improving Royal Mail's already highly valued public image and role in the community
- The need to sustain the centrality of quality to the business in the eyes of all employees, refocusing their efforts on various initiatives
- The value of benchmarking with organizations in different sectors on their performance, processes, tools, and techniques used (gaining knowledge)
- The need to reduce variations in the maturity of total quality in different parts of the organization
- The usefulness of sharing best practice within the organization to increase the overall standard of quality-enhancing activities
- The need to exact the maximum amount of information from the vast amounts of quality-related data collected from all over the organization, regarding performance and perceptions (converting information to knowledge is critical)
- The need to ensure that when processes cross functional boundaries they do so seamlessly

Royal Mail has a long history and tradition dating back to the 17th century. It faces a challenging future in the years to come with the pursuit of quality and business excellence.

COMPANY

United Way of Middle Tennessee

PRODUCT/SERVICE:

Raise and distribute resources to local nonprofit health and human service programs

LOCATION:

Nashville, Tennessee

NUMBER OF EMPLOYEES:

60

KEY PERSONNEL:

Case Study Contributor: Phil Orr

KEY USES OF INFORMATION:

- Customer satisfaction index
- Excellence in Service Quality Award feedback

ORGANIZING FRAMEWORK TO CREATE KNOWLEDGE:

Excellence in Service Quality Award criteria using the Plan-Do-Knowledge-Act model

AWARDS AND SIGNIFICANT RESULTS:

■ Excellence in Service Quality Award (ESQA) Bronze Level Award (1997)

■ Tennessee Quality Award—Commitment Level Recipient (1996, 1997)

INTRODUCTION AND BUSINESS OVERVIEW

United Way of Middle Tennessee (UWMT) consists of Davidson County (Nashville) and 13 regional partner counties. Davidson County represents 85% of the total campaign and has over one half of the population covered in the 14-county region. Founded in 1922 in Nashville, UWMT expanded into a regional operation in 1986. UWMT is a member of United Way of America and a member of the Tennessee Association of United Ways.

Governance and policy-making are provided through a volunteer board of directors, with day-to-day oversight and guidance provided by an executive committee of volunteer leaders. UWMT operates through an extensive group of volunteer committees built around major programs and services: resource development, fund distribution and community problem-solving, regional development, marketing and communications, financial management, information management, volunteer development and placement, workplace trends, human resources, quality process management and strategic planning.

UWMT at a Glance

Mission:	The mission of UWMT is to be the single most efficient way to identify, address and support solutions to local community health and human service needs.
Values:	UWMT is results-oriented and accountable, customer-focused, innovative, collaborative and inclusive, committed to integrity and caring.
Agencies Benefiting:	More than 450 agencies serving Middle Tennesseans received funding in 1998.

Dollars Raised in 1997:	$26,000,000 ($22,000,000 in Davidson County and $4,000,000 in regional counties)
Voluntary Effort:	More than 2,500 volunteers provide direction and resources to help United Way accomplish the mission and strategies.

UWMT historically has been viewed as the mechanism in the community to address its health and human service needs. It has been able to meet and balance the needs of multiple constituencies, including donors, agencies, employers, and the various needs of the community. Since 1990, however, the environment has been rapidly changing, and UWMT experienced difficulties, causing the need for strategic change. We identified the donor as our primary customer. Donor feedback, especially in our key accounts (94 employee campaigns raising $15,000 and up) and from leadership givers (individual donors giving $500 or more per year), indicated that "business as usual" was not acceptable.

A dual strategy to address these donor concerns was adopted. One leg of the strategy is to have a cost-effective, donor-oriented campaign. The donor was given the ability to designate his/her gift to any local health and human service agency or to the Community Solutions Fund. The Community Solutions Fund's dollars are allocated to programs with demonstrable outcomes that meet priority needs. The following customer requirements are having a major impact on UWMT.

Donor Requirements

1. **Trust/Accountability**
   - The 1992 national scandal raised doubts about the trustworthiness of United Ways. Since then, the United Way has faced intense scrutiny by local media, and general public trust levels are below pre-1992 levels.
   - Full disclosure and accountability are key organizational values and are important in regaining the high level of trust we seek with the public.
2. **Low Cost**
   - Donors want United Way to operate as efficiently as possible so that more money goes to help people.
3. **Choice**
   - Donors want to be able to direct how their gifts are used.
   - Having choices is important even if choices are not used.
4. **Results**
   - Donors want assurance that funded programs are helping people in need.

- Donors expect that their contribution is helping to solve local problems.
5. **Thanks/Recognition**
- Donors do not want elaborate or expensive recognition.
- Donors want their gift to be acknowledged, to be assured that the gift was used as intended and that it is appreciated.

Supplier Relationships

The primary supplier groups for UWMT are the agencies whose programs are funded through outcome-based investments. This funding is on a 2-year cycle that began in 1997. These program proposals demonstrate measurable client behavior or condition changes. These programs meet priority needs identified in the *1996 Nashville Needs Assessment* and support the outcome statements of the Community Solutions Councils. Funding decisions were approved by the board of trustees in late March 1997, and 184 programs at 80 agencies were funded.

Campaign Results Tell The Story

After extensive analysis of campaign results and trends, UWMT has focused on the following areas for the past two campaigns.

1. *Increasing the Davidson County campaign.* Upon analysis we determined that our annual growth was coming from our regional partner counties while Davidson County numbers were declining. With a plan to grow the Davidson County campaign, the 1996 Davidson County growth was 11% and the 1997 growth was 5%.
2. *Growing the Community Solutions Fund.* After several years of declining dollars to the Fund despite overall campaign growth, UWMT established a goal to raise 1 million new dollars for the CSF. UWMT exceeded this goal by raising 1.3 million new dollars.

Participation: Overall participation in the Davidson campaign had remained flat at 43% from 1994–1996. In 1997, a jump in participation to 45% (of the number of people solicited in workplace campaigns) was experienced.

Leadership Giving (donors giving at least $500 annually): While overall workplace participation is declining, workplace leadership campaigns have grown from 30% of total campaign dollars in 1991 to 40% in 1996. Although highly successful, they still represent a smaller group of donors. Of the key accounts, 54% of the dollars come from gifts of $500 or greater. Interestingly, this represents just 5.52% of the donors in these companies.

Designations/Donor Choice: In an effort to meet customer requirements, curtail competition, and lessen pressure on donors getting multiple requests, UWMT opened up the 1994 campaign to designations to all health and human service agencies providing local services. Since then it has seen a steady increase in donor designated gifts.

CONCLUSION

On March 29, 1995, the UWMT board of directors adopted a 5-year Strategic Plan, as presented by the senior team and the volunteer-led Strategic Planning Committee. Based on workplace and donor information, experience and research, the board set strategic directions and outlined tactics and action plans that will enable the organization to fulfill its mission.

UWMT has determined that its direction must be comprised of a dual strategy. This dual strategy is focused on a donor-oriented, cost-efficient workplace campaign *and* a dynamic allocations process designed to meet the primary needs of the community using aggressive problem-solving based on measurable outcomes and results-oriented spending.

Donors have told UWMT that they want to see how their dollars improve the community. In response, UWMT has designed an allocations process in which agencies develop program proposals that address defined community needs. Those programs must have measurable goals, and UWMT is measuring the results.

Finally, and most importantly, UWMT is promoting the Community Solutions Fund allocations process to donors as the best choice with the greatest positive impact for the community. Dollars designated to the Community Solutions Fund will be used for programs meeting the key deficits outlined in the *1996 Nashville Needs Assessment*. UWMT wants its donors to see the Community Solutions Fund as a profitable investment in the community's future.

PERFORMANCE EXCELLENCE JOURNEY

UWMT's quality journey began in 1993 with the appointment of a new President and CEO who was committed to transforming the organization based on quality improvement principles. This journey initially involved gaining the Board of Trustees support and providing that group with an orientation and overview of a process improvement plan. This orientation was provided by an Executive of Saturn Corporation, who was a Board member. Next a volunteer Board level committee was appointed to assist staff in its improvement efforts. This committee helped establish a quality journey plan, provided training resources and reviewed our assessment efforts. Next all staff were oriented in basic quality principles and senior managers were trained.

One of the challenges was to convince senior managers that the behaviors they model were critical to the success of the performance excellence journey. In short, a radical cultural change had to occur that was led by the President and the senior managers. UWMT relied on Deming's teachings as the foundation for this new approach to managing its business. First, the mission of UWMT was redefined and a 5-year strategic plan was created that was adopted in 1995. This vision was shared with stakeholders. Next UWMT taught managers that a foundation of trust with all employees had to be built. It began to eliminate the "blame" mentality and focused on process improvements. It discovered that people are more productive if system issues were dealt with, and if fear is reduced or eliminated. One example is feedback from employees indicated that the former computer system was very inadequate and was a source of frustration for many employees. UWMT submitted and received a technology grant from a local private foundation, and the technology issue has virtually been eliminated as a source of complaints. Finally, UWMT has focused on the needs of its customers, the contributors, and other stakeholders to continuously improve its business.

A Quality Council was formed in 1995. Its purpose is to promote continuous improvement throughout the organization and in all departments. The Quality Council is comprised of senior managers, process team leaders, and the director of research. The Quality Council meets monthly to review business plan goals, to monitor quality measurements and progress toward meeting quality improvement targets, to select organizational processes for analysis, to appoint team leaders and members, and to celebrate team successes.

The Quality Council in 1995 developed a list of the organizational processes and then prioritized those that were most in need of process analysis. The criteria for selecting the priority processes was based on customer survey feedback and managers' assessment of improvement needs. Annual mail surveys are sent to donors who contribute $500 or more per year and a general public random phone survey is conducted each year. In addition, campaign volunteers are surveyed and focus groups are conducted with the Loaned Executive group (campaign volunteers on loan from local organizations for 2 to 3 months to assist in the campaign) each year. From these surveys and focus groups it was clear that the pledge card needed to be redesigned (it was too confusing for many donors), public mailings were inefficient and resulted in some people receiving multiple copies of mailings, and the designations processing cycle time needed to be reduced in order to provide information needed to implement a redesigned fund distribution system. These processes were the first to be selected for analysis and improvement. Teams were appointed by the quality council and were trained by our local IBM representative using an IBM method, *Skill Dynamics: 10 Steps for Process Improvements.*

The Quality Council decided in 1995 to prepare the organization to apply for a national nonprofit quality award, the Excellence in Service Quality Award (ESQA), sponsored by United Way of America. This award is modeled after the Baldrige criteria. UWMT's first application was submitted in March of 1996; it also submitted an application for the Tennessee Quality Award the same year. The Excellence in Service Quality Award process involves both a written feedback report and a site visit discussion. The UWMT review team included Lee Matthews of the Tennessee Valley Authority. He volunteered to meet with the senior management team to present his system for integrating feedback into improvement opportunities. In 1996 UWMT had two written feedback reports (ESQA and Tennessee Quality Award) to utilize for improvement opportunities.

After the presentation from Mr. Matthews, the decision was made to use a modified version of the feedback integration process. The timing of this review process was coordinated to inform the management team's creation of the annual business plan that supports the organization's strategic plan. The flow chart (Figure 1) illustrates the steps in the feedback integration process.

1. Members of the management team are also members of the Quality Council. Beginning in 1998, UWMT alternated applying for the Excellence in Service Quality Award and the Tennessee Quality Award so that at least one feedback report will be received each year. The management team discusses each category feedback item in a meeting.
2. Customer survey data is reviewed throughout the year. However, during the feedback integration process this data is summarized and prioritized by the management team
3. The chart shown in Figure 2 is used by the management team to consolidate feedback into opportunity areas. All feedback items regardless of source are arranged under the appropriate Baldrige categories, Leadership, Strategic Planning, Customer Focus, Infor-

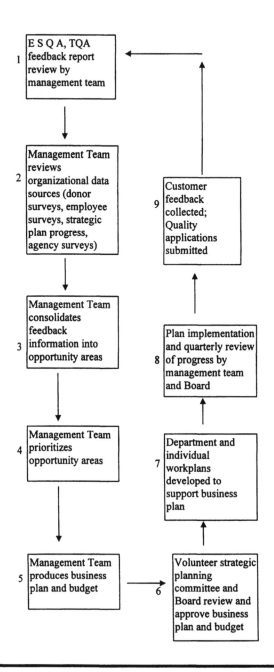

Figure 1   Feedback Integration Loop.

Category 1: Leadership Feedback Summary—Opportunity Areas

Comment area - Category - 1	We Have Heard This From:						Need for Improvement? (1 - 5 POINT SCALE)
	Quality Awards Feedback	Customer surveys	Donor Face-to-Face Feedback	Volunteer and Staff Feedback	Critical Organization Indicators	Remarks	
Not Clear how 5-year strategic plan is modified and improved as future opportunities arise							
Not Clear how senior management uses information to define and prioritize improvements							
How are conversations at all levels of organization regarding Senior Management Team issues captured and analyzed as part of planning process							
Not clear how data is used to track progress relative to plans and how use of information permits timely, effective decision making							
It is not evident how UWMT manages risks associated with managing its operations							
Are other managers besides the President actively involved in community organizations?							

Figure 2 Consolidated Feedback.

mation and Analysis, Human Resource Development and Management, Process Management, and Business Results.

4. This form is completed by each management team member. Scores for each item are totaled and averaged for the team. The items are then rank-ordered from high to low. A discussion of each item takes place with consideration given to priority from the score, the organizational capability to implement the improvement opportunity, and additional human or financial resources required for implementation. The team decides which items to work on for the coming calendar year. Typically any item with a score of 4 or higher on the priority list is worked on in the coming year. UWMT's relatively small staff size makes it easier for the management team to quickly assess resource capabilities and action steps necessary for improvement strategies.

5. The annual business plan is developed in October along with the budget by the management team after the opportunity areas have been prioritized.

6. The volunteer strategic planning committee reviews management's business plan proposal and reviews, amends the plan before final approval by the Board of Trustees.

7. Department Work Plans are developed directly from the business plan and inform individual work plans.

8. Periodic reviews of plan progress are made at the management team level and quarterly progress reports are shared with the Board and all staff.

9. The ongoing collection and analysis of customer feedback completes the feedback loop. UWMT applies for at least one Baldrige-based award each year in order to receive the feedback reports, which are also integral to the feedback integration process cycle.

In summary, the steps in the UWMT feedback integration process are straightforward and relatively simple in terms of approach and deployment. UWMT's relatively small staff size eliminates the need for more complex analysis of resource and deployment needs and strategies. On the other hand it may be somewhat unique in requiring volunteer committee review and approval of staff-generated plans and strategies.

RESULTS

The process of integrating feedback from various sources and integrating this information into the annual business and strategic planning process has enabled UWMT to focus on priority issues within the framework of the strategic plan. Two years of results reflect the implementation of the Community Solutions Council Process and the integration of assessment feedback with other organizational feedback tools. Shown below are trends in several key areas crucial to UWMT's success.

Total Campaign Growth and Strategic Goals

Year	Strategic Goal	Dollars Raised	Dollar Growth Over Previous Year	% Increase
1995	$22,206,280	$22,635,470	$1,483,000	7.0%
1996	$23,419,000	$25,006,000	$2,370,530	10.5%
1997	$25,153,662	$26,010,000	$1,004,000	4%

Since the implementation of UWMT's strategic plan in 1995 it has exceeded overall campaign growth each year. This amount includes designated gifts, donor restricted gifts and gifts to other United Ways, as well as unrestricted donations that go to the Community Solutions Fund.

Growth to the Community Solutions Fund (CSF)

Year	CSF Dollar Amount
1996	$6,300,000
1997	$7,600,000

The 1997 campaign was the first year since 1995 that the Community Solutions Fund showed an increase over the previous year. UWMT exceeded the goal to raise 1 million new dollars in the Fund.

General Public Overall Favorability

Year	Total Favorable Toward UWMT	Very Favorable	Total Unfavorable
1996	75%	34%	19%
1997	76%	40%	16%

While Total favorability (very and somewhat favorable) is up slightly the Very Favorable rating increased by six percentage points and the Total Unfavorable rating decreased by three percentage points.

General Public Trust

Year	High Trust
1996	63%
1997	71%

United Way's high trust level with the general public has improved significantly between 1996 and 1997.

Overhead

Year	Overhead Percent
1996	19.4%
1997	16.8%

10

HOW THREE WORLD-CLASS GOVERNMENT SERVICE ORGANIZATIONS IMPLEMENT PRINCIPLES OF PDKA— FEDERAL SUPPLY SERVICE NORTHEAST AND CARIBBEAN REGION, TENNESSEE VALLEY AUTHORITY'S FOSSIL & HYDRO POWER, AND NATIONAL AERONAUTICS AND SPACE ADMINISTRATION'S KENNEDY SPACE CENTER

INTRODUCTION

The companies that provided these case studies—The National Aeronautics and Space Administration Kennedy Space Center, the Fossil and Hydro

Power portion of the Tennessee Valley Authority, and the Federal Supply Service Northeast and Caribbean Region—vary in the services they provide. Each organization is a federal entity, but each has a different mission to accomplish. All three case study companies have achieved significant accomplishments toward their respective missions.

The National Aeronautics and Space Administration Kennedy Space Center

The National Aeronautics and Space Administration (NASA) is in the midst of a process of evolution and change, largely dictated by tight budgets, dwindling resources, and a shrinking work force. President Clinton and Vice President Gore have pointed to NASA as a prime example of their National Performance Review initiative to "reinvent government" by reducing bureaucracy, cutting costs, and improving efficiency. Emphasis has been placed on six areas: goal-setting, planning, reviewing quality and operational performance, communicating with employees and encouraging involvement, recognizing employee contributions, and customer focus. The John F. Kennedy Space Center (KSC) is a world leader in space launch, landing, and payload processing operations. All of the work involved in ensuring a successful Space Shuttle mission comes together at the Kennedy Space Center.

Amid the changes, the human elements of trust, ethical behavior, and a strong commitment to KSC's goals are prevalent characteristics of its employees. The pride with which it does its launch and landing job has not been compromised. In 5 years, KSC has reduced the cost per Shuttle flight by $164 million, more than 50%. During the same period, it reduced the total number of labor hours per Shuttle mission by 800,000, a reduction of 50%. These operational efficiencies were achieved through the efforts of hundreds of Continual Improvement teams.

Tennessee Valley Authority's Fossil & Hydro Power (F&HP)

The Tennessee Valley Authority's Fossil & Hydro Power's (F&HP) service is electricity. The Tennessee Valley Authority (TVA) is a federal agency, but it has many characteristics of a private corporation. Eighty percent of the electricity it generates is from fossil plants and from hydro units. These plants have been the mainstay of the agency since the 1950s. Because of its aging fossil plants and dams, the F&HP organization began its quality initiative to be more productive and focus on those things that would improve the bottom line. In the process, it learned how to take information from feedback reports and other sources, and it developed a strategy for implementing areas of opportunity into its core business. The case study details how a knowledge-based model has been developed and refined and explains what accomplishments have resulted.

TVA's F&HP realized that it needed to couple the information that it gathered with a theory in order to obtain usable organizational knowledge. In other words F&HP needed to convert its information, including its assessment feedback, into knowledge. By creating a four-step PDKA model, TVA's F&HP improved its bottom line significantly. F&HP has been a finalist for the President of the United States Quality Award, winner of the Tennessee Quality Award, and winner, finalist, and semifinalist for the Rochester Institute of Technology's USA Today Quality Cup. F&HP was also awarded the Hammer Award from Vice President of the United States Al Gore, and was Best-in-Class in financial and operational results.

Federal Supply Service, Northeast and Caribbean Region

Federal Supply Service, Northeast and Caribbean Region (FSS NECR) is a division of the U.S. General Services Administration. FSS NECR's key services are procurement and management of a wide range of office supplies, paper products, and packaging materials; quality assurance and contract administration of suppliers; and lease cars, trucks, and various special vehicles. FSS NECR's mission is to provide federal agencies worldwide with common use supplies and services necessary for success in their various program areas. Its mission also includes the transfer, donation, and sale of excess and surplus personal property assets of the federal government.

In a 3-year benchmarking study against the private sector, FSS NECR's prices were determined to be approximately 20% less than those of a major discounter of office suppliers and paper products. It continues to gauge itself against the best discounters and has found that its prices consistently average significantly below those of the private sector. FSS NECR has won the President's Quality Achievement Award, the President's Quality Improvement Prototype Award, and the New York Federal Executive Board's Total Quality Management Award, among others.

HOW THESE COMPANIES HAVE IMPLEMENTED THE PRINCIPLES OF PDKA

Consolidation of Information

KSC has a diverse group of customers and stakeholders in private industry, academia, and the public sector. Its principal external customers are the international science community; the astronauts who fly on the Shuttle;

other NASA centers whose hardware and software become a part of the mission; the media; Congress and the Executive Branch; local, state, and federal regulatory agencies; industry and academia; and ultimately, the American taxpayer. KSA's key uses of information are internal self-assessments using the President's Quality Award criteria; external assessments and town meetings; surveys and one-on-one interviews with employees; Florida's Sterling Award feedback; and benchmarking data. Each of its assessments had great value. The areas for improvement identified in each assessment were given a close review and steps were initiated to strengthen any weaknesses. The assessments also forced its line organizations to take a critical look at its quality efforts at all levels; educated managers and employees about the meaning and importance of quality; and provided senior management with an expert impartial, outside diagnosis of both their strengths and opportunities for improvement. The key types of data KSC uses include quality, efficiency, timeliness, customer satisfaction, safety, environmental, employee development, employee involvement, diversity, and supplier performance.

Effective information and analysis systems are required to enable KSC to meet its challenges with the highest levels of quality, cost, and schedule performance ever achieved at KSC. The information and analysis systems also enable outstanding successes in continual improvement, as reflected in the positive results. Many different methods are used to provide data with integrity in its information and analysis systems. KSC defines data integrity to include data accuracy, consistency, reliability, validity, and completeness. All data systems employ appropriate electronic security measures. Work documents, part inventory tags, and warehouse locations are bar-coded to enhance traceability, reduce data entry errors, and improve data timeliness. KSC comparative studies cover the full range of the benchmarking spectrum. Competitive comparisons and benchmarking studies are performed in most functional areas of KSC. KSC has a Benchmarking Network, which is a collaboration of NASA and nine major KSC contractor organizations. Together they have pioneered and customized consortium benchmarking techniques to enable more cost-effective benchmarking studies. The methodology of the KSC Benchmarking Network was tested with a pathfinder study of government property management, which has already resulted in the cost savings. Benchmarking studies are also being refined with external inputs, such as feedback from the President's Quality Award and Florida's Sterling Award applications. The information from the feedback reports has been incorporated into the KSC benchmarking strategy.

TVA's F&HP's key uses of information are state and national award assessments, employee surveys, customer surveys, and key business indicators. F&HP has taken feedback from five sources, the driver being the President's Award feedback report, and has developed a strategy for implementing areas of opportunity into its core business. F&HP forms an action plan team (steering committee) responsible for consolidating information sources into a feedback summary by integrating the Business Process Review feedback report with existing information sources in the organization, such as employee surveys, customer feedback, past assessments, and others as appropriate. They integrate all of this critical information by putting it into a matrix that uses the seven Baldrige categories as the organizing framework. This integration results in information being converted into knowledge.

FSS NECR maintains extensive databases containing detailed information on customers and their buying preferences. Market trends and customer purchasing patterns are closely monitored to ensure that products are offered at competitive prices and that delivery to customers is uninterrupted and prompt. FSS NECR has taken advantage of numerous new automated databases to ensure reliability, rapid access, and rapid update of data and information. Sales data relative to the products of the Commodity Center are updated regularly and available on-line through the Marketing Information System. A vast array of programmatic information is available on-line through systems such as FSS-19, the Fleet Management System (FMS) database, and the Sales Automated System (SASy). Improving the scope of information and data is important. FSS NECR continually examines its databases, data collection systems, and data updating processes to ensure the decisions, strategic directions, and adjustments it is making are based on the latest facts. FSS NECR selects competitive comparisons and performs benchmarking to improve performance. As a result of benchmarking and gaining critical information, FSS NECR is now achieving a high level of performance. The key operational performance indicators FSS NECR uses to monitor the delivery of services in its main program areas are Number of Customer Backorders, CSC Fill Rate, WDC Fill Rate, Stock Program Profit and Loss, Percentage of Stock Items Covered by Contract, Special Order Program Work in Progress, Dollar Value of Schedule Items Covered by Contract, Percentage of Quality Complaints Resolved on Time, Percentage of Contracts Delivered on Time, Percentage of Pre-award Plant Surveys Completed on Time, Fleet Utilization, and Surplus Property Transfer and Donations Completed on Time. FSS NECR collects data from customers on a daily basis. Customer Service Directors (CSDs) handle numerous calls each day for assistance in a wide variety of areas.

Another key means to determine customer requirements is FSS NECR's program of on-site customer visits. Here they meet face-to-face with customers and discuss their needs and level of satisfaction with FSS' products and services. FSS NECR segments its customers according to their delivery needs. Recent trends have indicated that more customers are interested in overnight delivery, a service not previously provided by FSS NECR. Identification of this customer requirement has led to the development of the FSS NECR "Next Day" contracts. Customer Information Seminars provide an excellent vehicle for FSS NECR to detect and then probe for changing requirements that may be in store for its customers. FSS NECR also detects future changes in customer requirements by studying new Executive Orders and legislation that might affect agencies' requirements.

An Organizing Framework

KSC's organizing framework to create knowledge is its information and analysis system. F&HP's organizing framework to create knowledge is the President of the United States Quality Award Criteria. FSS NECR uses its indicators to represent a comprehensive framework to provide appropriate information to ensure that these indicators also provide the information necessary for managers to determine where deficiencies are occurring and how proper adjustments can be made. It also uses The President of the United States Quality Award criteria as a way of organizing its initiatives.

Clear Decision-Making Criteria

At KSC, several criteria determine the data used for improving performance. For example, the data must (1) be effective in measuring desired performance and in forecasting results; (2) provide information to identify problems and corrective actions; and (3) be economical to collect. A KSC Measurement Handbook and Workbook was developed to assist process owners in planning and establishing their performance measurement programs. KSC's customer survey process provides a prioritized method for deciding which systems to improve. Administered by an objective group in the organization, each customer payload team is surveyed after each launch. The survey requests that the customer rate each applicable item in the survey on a scale of 1 (poor) to 5 (excellent). Any score below 4.0 becomes a priority candidate for improvement.

TVA's F&HP determines the importance of each strength and weakness area by dividing the areas into two matrices, one for areas of weakness and one for strengths. The purpose of these matrixes is to determine those areas of opportunities which, if improved, would help the organization the most. To assist in the prioritization of all opportunities, the action plan team uses a criteria by which to make decisions and measure priority. The six criteria are (1) potential impact on the customer (good or bad); (2) potential impact on the bottom line; (3) severity of the problem; (4) ability of a task team to solve; (5) ability of a task team to solve quickly; and (6) ability of a task team to solve with low cost. These six criteria are scored on a scale from 1 to 5, which allows each opportunity to be quantifiably weighted. Scores are totaled and priorities are determined based on which scores are higher. The highest scoring opportunities are then used by the Action Plan Team to determine how many of the opportunities can feasibly be done within given time and dollar constraints.

FSS NECR considers customer data and customer satisfaction to be extremely important, and many of its correlations to PDKA are in the customer discussions. FSS NECR aligns its measurement system with organization priorities. Examples of aligning types of data collected and the measurement of performance with respect to organization priorities are reflected by its key business drivers as follows: sales data supports the key business driver "maintain long-term profitability"; staffing and other personnel data meet the key driver "continue planning for operations in a downsizing environment"; market place data on competitors' prices help maintain a pricing structure better than the best offered by office supply discounters; and data on timeliness of delivery provide timely delivery including options that will satisfy or exceed the complete range of customer expectations. The criteria used for developing measures and selecting data for assessing quality and operational performance have their basis in the FSS NECR Strategic and Tactical Plans. Customer satisfaction is the goal at the heart of these plans.

Prioritization of Strengths and Weaknesses

TVA's F&HP states that information must be prioritized because all issues cannot be resolved at one time. The action plan team first stratifies all of the data to eliminate duplication, then the action plan team determines the priority of the strength and opportunity areas.

Assignment of Tasks

TVA's F&HP's action plan team assigns all selected opportunity areas as tasks that can be factored back into the business by using a six-step problem-solving process. The action plan team decides whether to form task teams. For those opportunity areas that will be assigned as tasks, the action plan team selects a sponsor, team leader, team members, and review team members.

Problem-Solving Process

TVA's F&HP implements improvements using a problem-solving process. Task teams are improvement teams whose members are appointed by management based on their expertise and are assigned a problem to work on. Each task team uses the problem-solving process to attack the root cause for each area of opportunity. The problem-solving process has the following steps: reason for improvement, problem definition, analysis, solutions, results, and process improvement.

Periodic Reviews

KSC uses the Associate Administrator Review Status (AARS) process to integrate financial and nonfinancial data. The AARS process is an organization-level review of data relating KSC's key health indicators (specifically in cost, schedule, and technical performance) to budgetary compliance and customer satisfaction. The AARS review includes discussions pertaining to all completed milestones and communicates any impact on KSC's progress to plan against its designated budget. The AARS also helps determine the budget estimate through fiscal year completion and its associated impact on the rest of NASA.

At KSC, the types of analyses performed on organization-level data are driven by user requirements and the types of data collected. Pareto, trend, cause-and-effect correlation, statistical process control, and root-cause analysis techniques are used extensively for organization-level data. Correlation between related performance measures also verifies various analysis results. Organization-level data reviews involve a combination of measures (cost, cycle time, quality, customer satisfaction, and others).

At TVA's F&HP, once strengths and opportunities have been integrated into business plans, management conducts Business Reviews, which include the status of the task teams and the implementation of the solution for the opportunity areas.

FSS NECR examines performance indicators to assess whether it is achieving the goals and objectives developed in its strategic planning sessions. When negative variance is detected, steps are taken to determine the root causes. Then an action or combination of actions, such as resource redeployment, process improvement, efforts to improve employee morale, training, analysis of special causes, and various other performance improvement techniques are taken. FSS NECR uses the vast array of data on competitive performance for operational reviews, planning, and decision-making.

Standardization

FSS NECR employees and managers seek out and demand data and facts to ensure that improvement priorities are properly aligned with key mission elements. They then measure results both to confirm that improvements had indeed taken place and to discover where more effort could produce even further improvements.

CONCLUSIONS

Strategic plans have been used to move NASA's Kennedy Space Center (KSC) toward the future. From these efforts, areas have been identified for review to ensure adequate coverage in the KSC plan and for consistency with the agency's documents. The plan encompasses short- and long-range goals. It considers (1) customer/supplier requirements; (2) work process changes from their continual improvement efforts; and (3) KSC's unique role in this nation's civilian space effort. KSC is confident that it will continue to change as its challenges dictate.

TVA's F&HP organization has successfully adopted the PDKA logic into its operations. It believes that it is critical to complete the information cycle and to implement solutions to problems, because these actions move businesses toward World Class.

FSS NECR uses information technology to gather and integrate its variety of information sources. This automation also allows easy access. Current levels and trends in operational and financial performance show extremely strong financial results. Their measures show positive trends and very strong results in all program areas.

COMPANY

Federal Supply Service, Northeast and Caribbean Region (FSS NECR) (A Division of the U.S. General Services Administration)

PRODUCT/SERVICE:

- Procurement and management of a wide range of office supplies, paper products, and packaging materials
- Quality assurance and contract administration of suppliers
- Lease cars, trucks, and various special vehicles

LOCATION:

New York City

NUMBER OF EMPLOYEES:

300

KEY PERSONNEL:

Case Study Contributor: Terrance Martin

KEY USES OF INFORMATION:

- *GSA Advantage!* System on Internet
- Maintain databases containing detailed information on customers and their buying preferences
- Customer information seminars, customer surveys, focus groups
- Alignment of measurement system with organization priorities
- Organizational assessments
- Multi-year trend data for key indicators
- "Schedules Information System"
- Competitive comparisons and benchmarks
- Linking information with processes

ORGANIZING FRAMEWORK TO CREATE KNOWLEDGE:

The President of the United States Quality Award criteria

AWARDS AND SIGNIFICANT RESULTS:

- The President's Quality Award Program Quality Achievement Award (1997)
- The President's Quality Award Program Quality Improvement Proto-type Award (1995)
- Vice President Al Gore's Reinventing Government "Hammer Award" (1994)
- New York Federal Executive Board's Total Quality Management Award (1993)
- Administrator's Quality Award (1991)
- President's Council on Management Improvement Award (1991)
- President's Council on Management Improvement Award (1989)
- Administrator's Productivity Improvement Award (1989)

INTRODUCTION AND BUSINESS OVERVIEW

The Federal Supply Service, Northeast and Caribbean Region (FSS NECR), is a component of the national Federal Supply Service, which is one of the major divisions of the U.S. General Services Administration. FSS NECR's mission is to provide federal agencies worldwide with common use supplies and services necessary for success in their various program areas. Its mission also includes the transfer, donation, and sale of excess and surplus personal property assets of the federal government.

FSS NECR's main offices are located in New York City and Boston. Its 300 employees, with an operating budget of $40 million, generated more than $800 million in sales in Fiscal Year 1996. This translated to an estimated $300 million in savings for taxpayers due to volume purchasing efficiencies and other value-added dimensions of FSS NECR's supply system, low vehicle lease rates, and the reuse of excess and surplus government personal property.

Employees include customer service and marketing specialists, contract specialists, quality assurance specialists, engineers and commodity management specialists, inventory managers, property disposal specialists, transportation operations specialists, automotive equipment repair inspectors and equipment specialists, clerical support personnel, data systems specialists, supply systems analysts, and managers. Educational levels run the full gamut from high-school graduates to Ph.D.s.

"Bargaining Unit" employees are represented by the American Federation of Government Employees (AFGE). Approximately 75% of FSS NECR's workforce is included in the "Bargaining Unit."

The FSS NECR organization includes the Office Supplies and Paper Products Commodity Center, the Contract Management Division, the Fleet and Personal Property Management Division, and the Customer Service and Marketing Division.

Office Supplies and Paper Products

The FSS NECR Office Supplies and Paper Products Commodity Center (Commodity Center), located in New York City, is responsible for the procurement and management of a wide range of office supplies, paper products, and packing and packaging materials used throughout the federal government worldwide. Key functions include identifying customer requirements, managing commodities, purchasing supplies through various contractual methods, managing inventories in the FSS national distribution system for resale to customers, and providing customer and vendor relation services.

The Stock Program entails contracting for and managing inventories of commodities that are stocked in the FSS national system of Wholesale Distribution Centers (WDCs) and retail Customer Supply Centers (CSCs). By consolidating the requirements of the entire federal government for common use items, Commodity Center Contract Specialists are able to obtain large volume discounts from suppliers. Customers place orders through the *GSA Advantage!* system on the Internet, or by phone, fax, or mail. CSCs carry the most popular products and ship within 24 hours of receipt of order. WDC and CSC product catalog prices include delivery costs and have been benchmarked consistently lower than those of nationally known office supply discount firms.

Through the Federal Supply Schedules Program, Contract Specialists negotiate contracts for commercial supplies with discounts that equal or exceed those granted by vendors to their largest customers. Agencies then issue orders against these contracts for delivery directly from the vendors. The advantage of this program is that agencies are able to acquire small or large quantities of commercial products at prices much lower than they would be able to obtain through individual contracts or orders.

Customer requirements for products not covered by the Stock and Schedule Programs are handled through the Special Order Program. Agencies submit requisitions to Commodity Center buyers who obtain competitive quotes and issue purchase orders for direct deliveries. This program enables customers to take advantage of FSS NECR's purchasing expertise in obtaining products that are not commonly used and cannot be economically stocked in the WDCs and CSCs.

Total annual sales for Stock, Schedules and Special Order Programs were $800 million for fiscal year 1996. Much of this sales volume represents purchases in support of FSS's socioeconomic and environmental objectives. $160 million worth of products from the National Industries for the Blind (NIB) were sold by FSS NECR in fiscal year 1996. Additionally, FSS NECR annually sells in excess of $200 million worth of products bought from small businesses, including over $15 million from minority-owned firms and over $15 million from women-owned firms. Sales of over 1,000 products with recycled content exceeded $120 million in FY 1996, up from $7 million just a few years earlier.

Contract Management

FSS NECR's Contract Management Division (CMD), headquartered in Boston and operating throughout the northeastern U.S. and the Caribbean, is responsible for quality assurance and contract administration for over 400 FSS suppliers. These vendors provide not only office supplies but also products managed by other Commodity Centers in the FSS national system.

Before contracts are awarded, Quality Assurance Specialists (QASs) carry out plant facility investigations to ascertain that suppliers will be able to meet contractual obligations. Once contracts are in effect, QASs monitor manufacturing processes to ensure that FSS customers consistently receive high quality products. Meanwhile, Administrative Contracting Officers (ACOs) ensure that suppliers comply with all terms and conditions of their contracts, including delivery schedules. When problems arise, ACOs arrange for prompt solutions or initiate contract termination procedures.

Fleet and Personal Property Management

FSS NECR's Fleet and Personal Property Management Division, with its headquarters in New York City, services customers requiring leased vehicles. The fleet consists of 14,000 conventional and alternate fuel vehicles and realizes over 150 million revenue miles annually. Federal agencies can lease cars, trucks, and various special vehicles from eight FSS NECR locations throughout the northeastern U.S. and the Caribbean at rates benchmarked significantly lower than the competition.

FSS NECR's Personal Property Program manages the utilization, donation, and sale of excess and surplus personal property of federal agencies in the northeastern U.S. Availability of excess government property is widely advertised through publications and an electronic bulletin board for use by other agencies. When excess property is no longer required for federal use, it is declared surplus and made available for donation to state-level agencies and qualifying nonprofit organizations.

Surplus property not selected through the donation process is offered for sale to the general public. Property seized or forfeited through federal law enforcement programs is also sold at FSS NECR auctions.

The FSS NECR Personal Property Program has offices in New York City and Boston and handles approximately $110 million worth of transactions annually.

Customer Service and Marketing

The FSS NECR marketing staff, based in New York City, maintains extensive databases containing detailed information on customers and their buying preferences. Market trends and customer purchasing patterns are closely monitored to ensure that products are offered at competitive prices and that delivery to customers is uninterrupted and prompt. The marketing staff also publishes newsletters and brochures that highlight the diverse product lines, encourage cost savings, and provide information on environmental requirements and other important issues.

Customer Service Directors (CSDs), based in New York and Boston, are responsible for all facets of customer relations in the northeastern U.S. and the Caribbean. CSDs handle telephone inquiries, visit customers, and regularly hold customer information seminars to enhance awareness and understanding of FSS programs. CSDs are also a key source of information for other FSS NECR employees about changing customer needs.

Quality History

FSS NECR's commitment to quality began in the late 1980s when the Office of Management and Budget began to promote Total Quality Management (TQM) with intensity. At that time FSS NECR was experiencing difficulties. Backorders exceeded 30,000, contractor delinquency rates were over 12%, and FSS NECR had a strained relationship with its major supplier, NIB. FSS NECR was perceived by its customers as a marginal "we know best" mandatory source.

Then FSS NECR senior managers attended numerous national conferences on TQM and subsequently integrated its principles fully throughout the FSS NECR organization. FSS NECR gave up its mandatory source status in favor of being allowed to recoup costs through markups.

FSS NECR developed its statements of Quality Policy, Vision, and Values and ensured that all employees understood these concepts. Numerous large group sessions, retreats, training courses, and open forums were held to ensure that the new culture became entrenched. Process improvement paradigms were developed and specific training in quality was given to all employees. Once equipped with an understanding of the FSS NECR quality ethic, employees began to focus on customer needs as opposed to rules and regulations, challenged every work process looking for improvements, and worked as teams generating an effective synergy. Employees and managers sought out and demanded data and facts to ensure that improvement priorities were properly aligned with key mission elements. They then measured results both to confirm that improvements had indeed taken place and to discover where more effort could produce even further improvements.

The list of awards shown signify significant milestones along this journey. Quality Management (QM) is now fully ingrained in all FSS NECR employees and is synonymous with the ordinary course of business. FSS NECR's winning of the 1995 President's Quality Improvement Prototype (QIP) Award reflects the degree to which quality has permeated its business.

Other Important Factors

As a nonmandatory source of supply for federal agencies, FSS NECR is in direct competition with major commercial supply discounters and vehicle leasing companies. Its success is directly related to its effectiveness in competing for market share against other suppliers. In order to excel in this challenging business environment, FSS NECR constantly strives to exceed customer expectations. Its strategy is to deliver value-added products and services that surpass those of its competitors. This allows it to generate new revenues in the shrinking federal market and increase savings for the agencies and the taxpayers. Focus on the customer, continuous improvement, and teamwork are key vehicles for achieving FSS NECR's goals.

FSS NECR receives no appropriations from Congress for its supply and procurement programs. All costs are recouped by a mark-up placed on the cost of goods sold. FSS NECR has relied heavily on new technology both to stay competitive and to handle an increasing workload with decreasing resources. All employees are equipped with computer workstations linked locally by a local area network (LAN) and to national databases by a wide area network (WAN).

All FSS NECR supply activities are governed by the Federal Acquisition Regulations (FAR) and related statutes.

Major New Thrusts and Future Challenges

FSS NECR has initiated a bold new venture, which flowed from strategy sessions following analysis of customer requirements and external factors. Customers have told FSS NECR that they want the option of overnight delivery of office supplies. The National Performance Review (NPR) stated: "Governance means setting priorities, then using the federal government's immense power to steer what happens in the private sector. As we reinvent the federal government, we, too, must rely more on market incentives and less on new programs." These factors led FSS NECR to develop and award its "Next Day" contracts. These contracts with five of the country's largest office supply discounters—Boise Cascade, Office Depot, Corporate Express, BT Ginn, and Staples, and Innovative Sales Brokers (a small disadvantaged business)—provide overnight delivery of a wide range of office products. This satisfies customer demand and aligns FSS NECR activities with the intent of the NPR. In the process, FSS NECR has formed partnerships with its former competitors for the benefit of the taxpayers.

> We will see positive change when government improves the way it delivers goods and services and when it stimulates competition by giving customers greater choices . . . NPR salutes FSS . . . for setting an example for positive change that others will surely follow."
> — **Robert Stone**, **Project Director of the National Performance Review**, in a letter written to the Senior Executive Service about FSS NECR.

To facilitate this new way of delivering products to its customers, FSS NECR has drawn upon its support suppliers—in this case its own Central Office—to allow ordering by Government Credit Card, which places no burden on FSS NECR's customers to establish accounts with "Next Day" suppliers. A simple credit card number is sufficient for FSS NECR's customers to use this FSS NECR "Next Day" contracts.

Management of Information and Data

Selection and management of information and data used for planning, management, and evaluation of overall performance. Listed below are the main types of data and information collected by FSS NECR:

Types of Data and Information Collected by FSS NECR

■ Sales data	■ Business line volume data
■ Customer satisfaction data	■ Customer complaint data
■ Financial performance data (profit and loss)	■ Staffing and other personnel data
■ Supplier performance data	■ Diversity data
■ Data on timeliness of delivery	■ Program performance data
■ Internal cost data	■ Program profile data
■ Market place data on competitors' prices	■ Data on participation and achievements in campaigns
■ Product quality data	■ Budget execution data
■ Training data	■ Rewards and recognition data

Alignment of measurement system with organization priorities. Examples of alignment of types of data collected and measurement of performance with respect to organization priorities are reflected by the key business drivers are given below:

- ■ *Sales data* are collected and analyzed to determine if business volumes will support the key business driver *maintain long-term profitability.*
- ■ *Staffing and other personnel data* are collected and analyzed to provide input into strategy development sessions necessary to meet the key business driver *continue planning for operations in a downsizing environment.*
- ■ *Market place data on competitors' prices* are collected and analyzed to assess performance relative to the key business drivers *maintain a pricing structure better than the best offered by*

office supply discounters, and *maintain a pricing structure for vehicles better than the best offered by commercial vehicle rental or leasing companies.*

■ *Data on timeliness of delivery* are collected and analyzed to assess performance relative to the key business driver *provide timely delivery including options which will satisfy or exceed the complete range of customer expectations.*

Reliability, rapid access, and rapid update of data. FSS NECR has taken advantage of numerous new automated data bases to ensure reliability, rapid access, and rapid update of data and information. For example, complete and highly detailed personnel information is available on-line through a Lotus Notes system. Access is restricted to only those few managers who have a need to know this information. Sales data relative to the products of the Commodity Center are updated regularly and available on-line through the Marketing Information System. A vast array of programmatic information is available on-line through systems such as FSS-19, the Fleet Management System (FMS) database, and the Sales Automated System (SASy). SASy gives information on personal property sales. Information such as inventory levels in FSS Wholesale Distribution Centers around the country is available on-line through FSS-19. The FMS provides complete data on use, assignment vehicle maintenance history, and other information concerning FSS NECR's 14,000 vehicle fleet. By virtue of the above-mentioned systems and other systems, data users always have reliable and up-to-date information and data.

The criteria used for developing measures and selecting data for assessing quality and operational performance have their basis in the FSS NECR Strategic and Tactical Plans. Customer satisfaction is the goal at the heart of these plans.

Through dialogue with customers and through other techniques FSS NECR has developed surveys which measure satisfaction in a range of key areas such as price, delivery time, quality, and customer service. Information from surveys is complemented with ongoing measurement of process output which directly impact customer satisfaction such as back-order rate, requisition fill rate, and vehicle cost per mile.

The assumption is that the delivery of quality products and services to customers is, to a substantial extent, dependent on the general quality of workplace culture. Thus, FSS NECR uses the annual GSA Quality Survey to measure the degree to which employees have embraced quality principles such as teamwork and the elimination of barriers to efficient customer service.

Improving the scope of information and data. FSS NECR continually examines its databases, data collection systems, and data updating processes to ensure the decisions, strategic directions, and adjustments it is making are based on the latest facts. FSS NECR evaluates its use of data in terms of its key priorities which flow from its key business drivers. For example, as FSS NECR examined its key business driver related to long-term profitability, that revealed that it had a significant deficiency in a key data area. Following the change to Industrial Funding for the FSS Schedules Program, it found that the information needed to ensure cash flow from these contracts was, at best, unreliable, sketchy, and incomplete. In the past, these data were used as a rough business indicator of the use of the FSS NECR schedules. No payments based on these data were required. Consequently, it was a low priority among the many other databases FSS NECR maintained. Now accurate and complete data are critical. Analysis showed that some vendors had been casual about reporting the volume of their sales to FSS NECR customers. FSS NECR had to quickly develop information relative to the Industrial Fund Fee (IFF) on the estimated 1,000 new contracts that were assigned to it.

The Boston Systems Staff was asked to create a database which would enable the FSS NECR Contract Administration Branch (CAB) to keep track of all contracts delegated to it and also provide a means of documenting contract actions which could be used to track problems and progress over the life of a contract.

The Systems Staff developed a sophisticated database that could not only load basic contract information on each delegation, but also maintain a running record of performance and "automatically" issue form letters at the request of the operator.

The new System was named the "Schedules Information System" (SIS) and was shared with all other Contract Management Zones nationwide. At the direction of the national Office of Quality and Contract Administration, the Boston office provided training in the use of SIS to the other Zones. The Systems Staff also made sure that the database was properly installed on the Regional LANs and that the ACOs were given "hands-on" training in its use. The system will ensure that FSS NECR is able to meet this new organizational priority regarding cash flow from its scheduled contracts.

Competitive Comparisons and Benchmarking

Selecting competitive comparisons and benchmarking to improve performance. A key business driver for FSS NECR is to maintain a pricing structure better than the best offered by office supply discounters. Based on this, pricing data are selected for competitive comparison purposes. The FSS NECR staff regularly analyzes competitors' catalogues to determine their pricing structures. This analysis of catalogues is then followed-up by calling competitors' outlets and determining discounts and other incentives that they might be offering. This combination of data is then used in the process of setting FSS NECR's pricing structure.

A key competitive comparison strategy is to examine pricing structure for bellwether items in order to ensure competitiveness. An example of this is the analysis and subsequent reduction of the price of copier paper. FSS NECR's assessment of pricing in April 1996 showed that it needed to reduce its selling price of this item. Accordingly, after determining how competitors were pricing this item, it reduced its selling price from $33.20 to $24.99. This aligned FSS NECR's pricing for this critical item with the key business driver referred to above. Additional analysis of competition and FSS NECR's financial posture led it to decide to reduce the price of this item again in October 1996 to $22.99.

Evaluating and improving the competitive comparison and benchmarking process. FSS NECR closely monitors its competition to detect any breakthroughs in process improvement, delivery times, product quality, ordering systems, and other key areas in order to focus on the key areas of its own operations that must be improved to stay competitive. The most significant example of this process was the analysis which led to the development and delivery of the "Next Day" contracts. As FSS NECR examined delivery time of its competitors, it became clear that the major office discounters had achieved a breakthrough in delivery time. They promised overnight delivery of office supplies. FSS NECR had not achieved this level of performance. Due to benchmarking and gaining critical information, FSS NECR is now achieving this extraordinary level of performance.

Analysis and Use of Organization-Level Data

Integration and analysis of data to support reviews, business decisions, and planning.
Listed below is a description of the key national operational performance indicators FSS NECR uses to monitor the delivery of services in its main program areas.

Office Supplies and Paper Products Commodity Center

- **Number of Customer Backorders** is a fundamental measurement of the effectiveness of FSS NECR's Stock Program. Continuous stock availability depends on well-coordinated procurement and inventory management. (Goal: less than 1% of orders received)

- **CSC Fill Rate** measures the order fill rate for quick delivery requisitions. Since customers are increasingly depending on FSS NECR to be their just-in-time supplier, this indicator is crucial. (Goal: 98%)
- **WDC Fill Rate** measures the percentage of customer orders that are filled on the first request. It is a critical indicator of FSS NECR's ability to meet customer requirements in a timely fashion. (Goal: 90%)
- **Stock Program Profit and Loss** is a critical financial indicator which provides FSS NECR with information on its performance in terms of sales margins and business volumes. It serves as an overall indicator of the Stock Program's financial strength and facilitates the determination of future pricing and margins. (Goal: Break even)
- **% Stock Items Covered by Contract** is a measure that indicates the extent to which FSS NECR has contracts in place for the placement of orders to replenish Wholesale Distribution Center (WDC) inventories. Lack of contract coverage eventually leads to backorders, decreased customer satisfaction, and loss of revenue. (Goal: 96%)
- **Special Order Program Work in Progress** indicates how long it takes FSS to buy relatively low demand items for agencies. In addition to measuring timeliness of service delivery, it records workload levels, which are examined by management to make resource allocation decisions. (Goal: 45 days or less)
- **Dollar Value of Schedule Items Covered by Contract** measures FSS NECR's performance in providing contracts that other agencies can use to order their supplies directly from vendors. FSS NECR's contracts offer very attractive prices and give customer agencies a lower cost option to buying on the open market. (Goal: 97%)

Contract Management Division

- **% Quality Complaints Resolved on Time** provides critical information on the timeliness and effectiveness of FSS NECR's procedures for resolving all customer complaints. (Goal: 95%)
- **% Contracts Delivered on Time** measures how frequently contractors fill FSS NECR's orders for replenishment of WDC inventories within the delivery terms of their contracts. In short, it reflects the quality level of FSS NECR's vendor base. (Goal: 95%)
- **% Pre-award Plant Surveys Completed on Time** is a critical indicator which tracks how well FSS NECR's Contract Management Division is serving the buyers in its Commodity Center who are its internal customers. FSS NECR's Quality Assurance Specialists must provide Contracting Officers with judgments on whether or not potential contract awardees are capable of producing and

delivering quality supplies on a timely basis. Timely completion of these surveys prevents delays in contract awards which, in turn, could lead to either inventory shortages or—when orders are placed to prevent such shortages—inflated stock levels. (Goal: 95%)

Fleet and Personal Property Management Division

- **Fleet Utilization** is the percentage of total FSS NECR fleet vehicles that are in active service. Vehicles in inactive status include those just received but not yet assigned to a customer and those turned in by customers for replacement but not yet sold. Thus, this indicator reflects how effectively FSS utilized its fleet assets. (Goal: 90%)
- **Surplus Property Transfer and Donations Completed on Time** measures how quickly FSS NECR can transfer or donate government assets which still have a useful life from one organization to another. (Goal: 95%)

The indicators listed above, along with related indicators, represent a comprehensive database and analysis framework, which provides FSS NECR management with the information it needs to ensure that its performance in all program areas is proceeding according to plan. These indicators also provide the information necessary for managers to determine where deficiencies are occurring and how proper adjustments can be made.

In addition to the above operations indicators, FSS NECR gathers a wide range of human resources information in such areas as staffing, Equal Employment Opportunity accomplishments, training, participation in quality management projects, performance management, and rewards and recognition. These data are examined in relation to operations data in order to determine if employees have the training and resources necessary to accomplish the overall organizational goal of providing quality service to customers.

Customers. FSS NECR collects data from customers on a daily basis. Customer Service Directors (CSDs) handle numerous calls each day for assistance in a wide variety of areas. From analysis of these individual transactions, FSS NECR can infer more broadly based knowledge of its customers and what their up-to-the-minute requirements are. CSDs compile this information and provide it to the appropriate program area for analysis and planning purposes.

An example of using this type of data relates to FSS NECR's order-taking capabilities which have improved dramatically in the last 2 years. Customers are eager to use FSS NECR's improved systems. However, the level of technological development varies greatly among FSS NECR's customers. Many of them have the Government Credit Card and want to use it to order from FSS NECR. At this juncture, however, Internet capability is required for customers who wish to order from the Wholesale Distribution Centers using the credit card. To fill this temporary void, i.e., customers who have credit cards but no Internet capability, FSS NECR has taken on the workload to meet this customer requirement. Its planning and analysis process led it to conclude that devoting these resources on a temporary basis was necessary to fulfill its key business driver commitment to provide easy ordering systems.

Data on customer ordering patterns is examined regularly through analysis of the Marketing Information System (MIS). Staff from the Customer Service and Marketing Division (CSMD) provides this information to the appropriate program area for analysis, planning, and decision-making purposes. The system includes analysis of customer ordering patterns. A recent analysis of a major military customer in New York State indicated a severe decline in business. Further research showed that this military installation's Self-Service Supply Center (SSSC) had closed. FSS NECR had been the source on inventory for the SSSC. The CSMD staff went to the site immediately and found that a private business had opened up in close proximity to the military installation and the former SSSC customers had migrated to this new business. Upon further examination, FSS NECR discovered that it could easily beat this new business' prices and could offer conformity with all the socio-economic objectives and environmental objectives of the procurement preference system mandated by law. The new business was not able to offer its customers this conformance. The CSMD staff then proceeded to sign up over 600 individual customers to use the FSS Customer Supply Center, which draws its inventory from FSS NECR.

Operational performance and organization capabilities. FSS NECR examines performance indicators to assess whether or not it is achieving the goals and objectives developed in its strategic planning sessions. When negative variance is detected, steps are taken to determine the root causes. Then an action or combination of actions, such as resource re-deployment, process improvement, efforts to improve employee morale, training, analysis of special causes, and various other performance improvement techniques are taken.

The "Next Day" contracts flowed in part from an analysis of organization capabilities which showed that FSS NECR could not meet the evolving requirement for overnight delivery of office supplies. Now that FSS NECR has awarded the "Next Day" contracts, it has this capability.

Analysis showed that FSS NECR needed a better system to reduce defect levels. This analysis led to the development and deployment of Statistical Process Control.

Analysis of future human resource requirements in the rapidly downsizing organization, along with an analysis of training history of current employees likely to be candidates who would migrate from non-professional job series to professional job series, revealed a significant need to increase training and development opportunities for these employees. This was accomplished, and now over 40% of FSS NECR Contract Specialist positions are staffed with former clerical employees.

To fill the void for clerical functions created by the wholesale movement of secretaries and clerks into professional positions, FSS NECR carefully examined its need for clerical services. This analysis revealed that the proper use of technology could eliminate the need for many clerical and secretarial positions. This has happened. The Office of the Assistant Regional Administrator, FSS NECR, took the lead by cutting its clerical staff in half (from four to two). Other offices followed suit.

Competitive performance. In addition to several other references made in this report on analysis of competitive performance in such areas as delivery capabilities, pricing, and ordering systems, FSS NECR is using the vast array of data on competitive performance compiled as a result of the Federal Operations Review Model (FORM) analyses. Information and data in these reports include comparisons to the private sector in many areas. FSS NECR uses these data for operational reviews, planning, and decision making.

Categories of Competitive Performance Data Collected and Analyzed

■ Cost to issue a contract	■ Marketing costs as a percent of sales
■ Cost per mile for vehicle leases	■ Customer complaint resolution cycle time
■ Stock turn	■ Sales per contract person-year
■ Delivery time	■ Order processing cycle time
■ Number of contracts issued per person-year	■ Inventory cost as a percent of sales

Relationship of customer data, improvements in product/service quality and operational performance to changes in financial indicators. FSS NECR has experienced extremely strong financial performance. Because of this, it has chosen to reduce prices on a wide variety of items. Some components of this strategy have already been developed and deployed, such as the reduction of copier paper price.

FSS NECR's bold venture into the "Next Day" contracts represents a major decision to enhance return on investment relative to human resources. These contracts will provide the entire federal government with overnight delivery of office supplies. The resources required to put these contracts in place represent a tiny fraction of what it costs FSS NECR to provide its office supplies and paper product through its traditional system of Wholesale Distribution Centers and Customer Supply Centers. The potential for saving human resources here is enormous. FSS NECR is following the principles of the National Performance Review and is exploiting the commercial distribution network already in the private sector. By using its huge buying leverage, it was able to produce extremely attractive prices on these new contracts. Use of this service by FSS NECR customers will reduce national FSS costs in distribution significantly.

Evaluating and improving processes.

Improving production processes. A major step forward to maximize supplier performance and improve product quality was taken when FSS NECR began focusing on defect prevention instead of defect detection. This was achieved when FSS NECR introduced Statistical Process Control (SPC).

Improving delivery times. FSS NECR realized that the time frames of its product delivery system were not satisfying the needs of all customers. The services offered through its contracting processes were then significantly improved through the introduction of "Next Day" contracts.

Improving ease of service process. FSS NECR assessed its vehicle drop-off services and found that some customer needs were not being met. The process was cumbersome for some customers not located close to one of its Fleet Management Centers. To improve this service it established additional drop-off sites greatly reducing the time customers had to spend returning FSS NECR vehicles.

Improving service delivery. A key function FSS NECR provides in its property management program is the sale of surplus government property. In the era of downsizing, it had to make significant changes to maintain this process. One of FSS NECR's key business drivers is to continue to operate in a downsizing environment. With fewer federal employees available to maintain its process as it sustained a 24% reduction in staffing, and as it anticipates further staffing reductions, it looked to the commercial sector for auction support. It contracted with commercial auction houses to conduct the sales that had formerly been done by its own staff. This had proved to be highly successful and very popular with FSS NECR customers. The process was improved by selling more vehicles at each individual sale.

Improving return on investment. A particularly bold move taken by FSS NECR in this area was the decision to sell in New Jersey vehicles returned in the Caribbean. Due to taxes, excessive paperwork, and other burdensome requirements in the Caribbean, the process of selling vehicles there was cumbersome and slow. FSS NECR analyzed the situation and determined that vehicles could be shipped to the continental U.S. and to the auction site in New Jersey for $375.00 per vehicle. Analysis of FSS NECR returns on sales of vehicles in New Jersey compared to that which it was realizing in the Caribbean, showed that it would recoup more than the $375.00 shipping cost as a result of its change of sale venue to New Jersey. The process has become more profitable for FSS NECR and more convenient for its customers.

Similar processes, along with appropriate measurements, are in place for all other production and delivery processes FSS NECR performs.

Organization Operational and Financial Results

Current levels and trends in operational and financial performance. The measures show extremely strong financial results. FSS NECR exceeded its planned target for net revenue set by its Central Office by over $1.1 million in FY 94 and by more than $16.9 million in FY 95. The Commodity Center is currently ahead of planned levels of net income by over $11.7 million.

The measures continue to show positive trends and very strong results in all program areas. FSS NECR's backorder rate of 3,796 is far below the Central Office target of 13,500. This represents a dramatic 82% reduction in the backorder figure existing before the introduction of Quality Management in 1988.

Two of the most important measures of FSS NECR performance, WDC Fill rate (system) and Essential Business Item (EBI) Stock Coverage rate are at extraordinary high levels. The WDC Fill Rate (system) indicates what product availability rate is being achieved through the procurement and inventory management operations of FSS NECR. This rate of 96.6% exceeds the Central Office target of 93%. EBI's are a subset of the Stock Program. These are our highest selling items most in demand by FSS NECR's customers. The current coverage rate is at 98% meeting the Central Office target.

In a three-year benchmarking study against the private sector, FSS NECR's prices were determined to be approximately 20% less than those of a major discounter of office suppliers and paper products. FSS NECR continues to gauge itself against the best discounters and has found that its prices consistently average 37% below those of the private sector.

FSS NECR has repeatedly benchmarked costs associated with its Fleet Program against industry. In every case it was found to be less expensive than its competitors, thus validating the cost effectiveness of our program. Competitiveness in our Personal Property Program is apparent in an overhead rate for auction sales services of approximately 7%. This is far below the industry norm.

Customer Knowledge

Determining Current and Near-Term Customer Requirements and Expectations.

FSS NECR determines its customers' near-term requirements in a wide variety of ways. Very important among these is its program of Customer Information Seminars conducted by FSS NECR's cadre of Customer Service Directors (CSDs). These seminars are two-way communication vehicles that offer the customers information about FSS NECR's programs and seek information from customers on their current needs. FSS NECR also seeks feedback on its performance. These seminars are conducted throughout the northeastern U.S. and in the Caribbean. At each seminar an evaluation form is distributed which is used to collect data and information on the quality of the seminar presentation itself, the quality of FSS' products and services overall, and on customers' current requirements.

Another key means to determine customer requirements is FSS' program of on-site customer visits. Here CSDs meet face-to-face with customers and discuss their needs and level of satisfaction with FSS' products and services. This system offers a comprehensive way of obtaining customer knowledge.

FSS NECR system of determining customers' near-term and longer term requirements is supported by these regular customer visits. They illustrate the day-to-day operations of the Customer Service Director Program and the operations of the Office of the Director of Customer and Vendor Relations in the Office Supplies and Paper Products Commodity Center.

FSS NECR also uses customer surveys, focus groups, user panels, feedback from *CustomerGram* articles that solicit comments from customers, analyses of individual transactions to determine trends in customer requirements, and interaction with customers' current requirements.

FSS NECR has maintained detailed information on its customer groups over the course of many years. Customers continue to fall into two broad groups: civilian agencies and military installations. FSS NECR has closely analyzed trends in light of military downsizing and found that the impact has been small and that military activities remain our biggest customer group.

FSS NECR also segments its customers according to their delivery needs. Recent trends have indicated that more customers are interested in overnight delivery, a service not previously provided by FSS NECR. Identification of this customer requirement has led to the development of the FSS NECR "Next Day" contracts. Additionally customers have told FSS NECR that they want more management information concerning the ordering patterns of their own agencies. The "Next Day" contracts were designed to accommodate this requirement.

Analysis of delivery needs has also led FSS NECR to identify customers who need copy paper delivered on a delivery schedule basis as a distinct subgroup. Since copy paper is by far the largest selling FSS NECR item, the importance of identifying this subgroup of customers and satisfying their needs is clear. FSS NECR established a special unit in its Inventory and Requisition Management Division to handle the needs of this customer group. Customer reaction has been very positive.

FSS NECR also segments its customer by product line. For example, fleet customers are treated as a distinct subgroup. Customer needs and requirements are identified and analyzed through use of the Fleet Management System database and through numerous face-to-face encounters with customers including Interagency Motor Equipment Advisory Committee (IMEAC) meetings and national conferences, and through information gleaned from day-to-day business transactions, e.g., the return of a fleet vehicle and related assignment of a new vehicle.

FSS NECR is in continual contact with its customers concerning their use of other suppliers. In these exchanges FSS NECR seeks to gain from the customer what needs they have that FSS NECR is not fulfilling. This information is analyzed and used as appropriate for strategic planning.

FSS NECR determines which specific product and service features are important to customers in several ways. Certain features are required by law and various Executive Orders. For example, the recycled content of FSS NECR's paper products must conform to the requirements of the Resource Conservation and Recovery Act (RCRA) and Executive Order 12873. Currently FSS NECR offers paper products with a minimum of 20% post-consumer materials (PCM) and 50% overall recycled content. FSS NECR works with its customers to help them understand why its products have certain mandatory features.

Desired product and services features are also determined by listening to customers express their needs and desires (a leading indicator) and through analysis of sales patterns (a lagging indicator). For example, the evolution of FSS NECR's "Next Day" contracts followed many specific requests from customers for overnight delivery. Regular monitoring of the Commerce Business Daily revealed that agencies such as IRS and NASA were soliciting for next day delivery of office suppliers. FSS NECR also checked its database on sales trends to see if in reality customers were migrating to quicker delivery options. These data did show growth of sales in FSS NECR Customer Supply Centers, indicating a growing preference for quick delivery.

FSS NECR uses its marketing Information System (MIS) to track gains and losses of customers. This is done on a monthly basis. Complaint analysis is also done on a monthly basis. Internally, FSS NECR listens closely to its employees on issues concerning the work environment, health, and safety.

As a result of the various techniques FSS NECR uses to ensure both broad and in-depth customer knowledge, it has concluded that the most fundamental customer requirements include timely delivery of value-priced products of high quality. Additionally, customers want ease of ordering and an efficient and effective complaint resolution and inquiry process which gives them quick answers to their questions. They also want access to more in-depth documentation that they can use as needed to supplement responses to individual questions.

Addressing Future Requirements and Expectations of Customers. Key Listening And Learning Strategies Used.

In making assessments of customers' future needs FSS NECR places great emphasis on listening carefully to its customers. Customer contact staff have received extensive training in listening techniques. CSDs file reports on customer visits after they occur. They are trained to ensure that they understand the details of what the customer is saying and to record the customers' needs in such a way that the rest of the FSS NECR organization can understand and respond to those needs when asked.

Customer Information Seminars provide an excellent vehicle for FSS NECR to detect and then probe for changing requirements that may be in store for its customers.

FSS NECR also detects future changes in customer requirements by studying new Executive Orders and legislation which might affect agencies requirements. For example, Executive Order 12873 had a direct impact on agency requirements for recycled products. At present FSS NECR is closely monitoring development such as agencies' increased use of paper products made without chlorine bleaching. If this trend continues, it will translate into an increase in demand from agencies for this type of product and a decrease in demand relative to the products from which the agencies are migrating.

These types of changes also include greater use of alternate fuel vehicles. FSS NECR is closely monitoring the availability of this type of vehicle, the state of the refueling infrastructure necessary to support alternate fuel vehicles, and general customer acceptance of these new products.

FSS NECR always asks for "early notice" when dealing with its personal property disposal customers. FSS NECR asks customers at seminars and conferences, and in all other dealings with them, to let it know as early as possible of impending moves or equipment purchases, which will create excess property that will be referred to it for disposal.

FSS NECR's contacts with military installations that are downsizing are an integral part of its customer knowledge system. The military has a network of Self-Service Supply Centers (SSSCs). Many of them are being closed as result of the general downsizing. FSS NECR's representatives have met with officials from several of these installations to inform them of the availability of similar supply service throughout the FSS network of Customer Supply Centers. FSS NECR's representatives explained the FSS program, assisted the military in establishing accounts with the CSCs, and offered on-going assistance from all FSS supply programs to fill the void created by the closing of their SSSCs.

The FSS NECR customer listening system makes it possible to address customers' future needs early on. The Defense Commissary Agency (DECA) had very demanding delivery requirements for a group of products. Delivery times were actually limited to "half-day" windows on specific days. Working closely with the customer and the FSS NECR supplier, FSS NECR was able to meet the stringent requirements of this customer and fulfill the delivery terms as planned. The close contact and other customer relations techniques worked to make this project a success.

Evaluating and Improving Processes for Determining Customer Requirements, Expectations, and Preferences.

After each Customer Information Seminar, FSS NECR managers review the seminar evaluation forms to determine the level of satisfaction attendees expressed concerning the seminar itself and with FSS NECR's products and services. The system involves looking for patterns and trends as well as for individual concerns that to be addressed. When a trend is established, such as "not enough time devoted to how to order," this information is noted and conveyed to the FSS NECR's planning group and analysis is done to determine overall course content, time available, relative importance of topics covered, and assessment of needs and requirements put forth by other customers. Then a judgment is made as to how the seminar should be modified, and assignments are given to subject matter experts to make any changes deemed necessary.

Improving the process. In 1995 and 1996, FSS undertook a major assessment of its entire Supply and Procurement (S&P) Business Line (B/L). This analysis, referred to as the Federal Operations Review Model (FORM), looked at all aspects of the S&P B/L. The Deputy Assistant Regional Administrator for FSS NECR served as a member of that team. Detailed customer satisfaction surveys which relate directly to FSS NECR's Office Supplies and Paper Products Commodity Center were carried out as part of the FORM process. Surveys included questions on level of customer satisfaction, intent to continue to do business with FSS NECR, relative importance of services and products provided, what the organization does particularly well, what the organization might do better, and additional products and services that could be offered that would meet customers' needs. Analysis of all these factors has been very important in evaluating and improving FSS NECR's processes for determining customer satisfaction.

COMPANY

Tennessee Valley Authority Fossil & Hydro Power Organization

PRODUCT/SERVICE:

Electricity

LOCATION:

Chattanooga, Tennessee

NUMBER OF EMPLOYEES:

5,383

KEY PERSONNEL:

Case Study Contributor: Monte Lee Matthews

KEY USES OF INFORMATION:

- State and National Award Assessments
- Employee Surveys
- Customer Surveys
- Key Business Indicators

ORGANIZING FRAMEWORK TO CREATE KNOWLEDGE:

President of the United States Quality Award Criteria

AWARDS & SIGNIFICANT RESULTS:

- Finalist, President of the United States Quality Award
- Winner of Tennessee Quality Award
- Winner, Finalist, and Semi-finalist, Rochester Institute of Technology USA Today Quality Cup
- Awarded the Hammer Award from Vice President of the United States Al Gore
- Best-in-Class financial and operational results

INTRODUCTION AND BUSINESS OVERVIEW

The Tennessee Valley Authority is a federal corporation, the nation's largest electric-power producer, a regional economic development agency, and a national center for environmental research. TVA's power-service area covers 80,000 square miles in the southeastern U.S., including most of Tennessee and parts of Mississippi, Kentucky, Alabama, Georgia, North Carolina, and Virginia. TVA also manages the Tennessee River, the nation's fifth-largest river system. TVA operates 11 coal-fired plants, three nuclear plants, 29 hydroelectric dams, and a pumped storage plant. Together, they provide 28,417 megawatts of net winter dependable generating capacity. TVA provides power to 159 municipal and cooperative power distributors, and directly serves about 67 federal and industrial customers in the Valley through a network of 17,000 miles of transmission lines. TVA supplies the energy needs of nearly 8 million people.

This case study presents how a United States Presidential Award Finalist has taken information from feedback reports and other sources and has developed a strategy for implementing areas of opportunity into its core business. This case study details how a knowledge-based model has been developed and refined over the past few years and explains what accomplishments have resulted.

Deming's Perspective of Knowledge

To put the power of knowledge in perspective, here is Dr. W. Edwards Deming's answer to a question posed to him in a seminar (from *The New Economics for Industry, Government, and Education*):

> **Question:** Please elaborate on your statement that profound knowledge comes from outside the system. Aren't the people in the system the only ones that know what is happening and why?

> **Answer:** The people that work in any organization know what they are doing, but they will not by themselves learn a better way. Their best efforts and hard work dig deeper the pit that they are working in. Their best efforts and hard work do not provide an outside view of the organization.

> The theory of knowledge teaches us that a statement, if it conveys knowledge, predicts future outcome, with risk of being wrong, and that it fits without failure observations of the past. . . . Information is not knowledge. Information, no matter how complete and speedy, is not knowledge. Knowledge has temporal spread. Knowledge comes from theory. Without theory, there is no way to use the information that comes to us on the instant. A dictionary contains information, but not knowledge. A dictionary is useful. I use a dictionary frequently when at my desk, but the dictionary will not prepare a paragraph nor criticize it.

As Dr. Deming states, F&HP needed to couple the information that it gathered with theory in order to obtain usable organizational knowledge. In other words F&HP needed to convert its information, including its assessment feedback, into knowledge. F&HP has found that when it combines information with theory, it creates knowledge—knowledge that has moved and will continue to move its business into the future.

Using Knowledge to Drive Improvement

For many years F&HP had not figured out how to take the information from Baldrige-style assessments, employee surveys, or customer feedback and use it to actually make improvements for its business.

The Old Approach

One approach that F&HP used in trying to make improvements was to take each area of opportunity and develop action plans for all of the opportunities. It created a mini-crisis, got everybody riled up, formed committees and task teams, and spent a lot of effort to try to accomplish everything. A lot of time and money was spent on those well-intended efforts, but little was implemented, thus few improvements to the bottom line were seen. In other words, F&HP lacked priority and a disciplined follow-through. The reason for the failure is that such an approach is too diluted and too uncoordinated. It lacked an integration of all of information and it lacked a clear way to prioritize the mass of information.

The PDKA Model Approach

After years of attempts, F&HP introduced a problem-solving process and tried a better improvement planning approach that converted all identified opportunities into knowledge, prioritized the knowledge before proceeding to develop action plans, then acted to deploy the solutions to these opportunities into the business through problem-solving teams. With this approach, the organization selected only the most important areas of opportunities to focus on. These areas of opportunity are linked to F&HP's problem-solving process and integrated into F&HP's business planning process.

How to Drive Improvement

This case study describes how a Presidential Award Finalist has taken feedback from five sources, the driver being the President's Award feedback report, and has developed a strategy for implementing areas of opportunity into its core business. F&HP was not good at doing anything significant with the problems it identified. It needed to convert *information* which was in the form of feedback into knowledge. It *integrated* all of these problems into its business processes, and then *acted* to deploy the solutions to these problems into its business. In other words, it concentrated on step 3 (converting information to Knowledge) and step 4 (Acting by deploying solutions and tracking results). The following charts the steps 3 and 4 of this process, and the following narrative explaining the steps of the process. Steps in the process preceded by an "A" are actions; those preceded by a "D" are decisions.

Step A1, Review findings from the feedback report: The Management Team reviews the feedback report, and accepts its use in improving the business. The report addresses pluses and minuses within the organization. This report is taken seriously; the management team does not rationalize why certain comments are as they are.

Step D1, Concur?: Managers must review the findings from the feedback report and make sure that they indeed understand and concur with the findings.

Step A2, Modify as appropriate: Once managers clearly understand this feedback report and have no need for modification or clarification of improvement opportunities, the report is considered final.

Step A3, Hand off the report: Then the feedback report is handed off to an action plan team to develop detailed action plans around the feedback in this document.

Step A4, Review the feedback report: The action plan team members (both managers and employees) review the feedback report to make sure that they understand the content.

Step A5, Consolidate feedback resources and integrate them with other sources: The action plan team (steering committee) is responsible for consolidating information sources (Figures 1 and 2) into a feedback summary by integrating the Business Process Review feedback report with existing information sources in their organization, such as employee surveys, customer feedback, past assessments, and others as appropriate. They integrate all of this critical information by putting it into a matrix which uses the seven Baldrige categories as the organizing framework. The categories at that time were: Leadership, Information and Analysis, Strategic Quality Planning, Human Resource Development, Management of Process Quality, Quality and Operational Results, and Customer Focus and Satisfaction.

At this point information is being integrated with theory (the seven Baldrige criteria) to achieve a new level of knowledge. Then they consider each piece of information (with some knowledge added) as either a strength area or an area of weakness.

They determine the importance of each strength and weakness area by dividing the areas into two matrices, one for areas of weakness and one for strengths. The purpose of these matrixes is to determine those areas of opportunities which, if improved, would help the business the

Consolidate information sources

Example

Feedback summary – strength areas

Comment area Category 1.0	Information sources					Remarks	Need for improvement?
	Federal Quality Institute	Total quality assessment	Site visit	Employees	Employees internal assessment		
Senior managers have researched (benchmarked), developed, and are actively leading the deployment throughout the organization	X	X	X	X	X		5
Managers throughout the organization participate in communicating and reinforcing quality values	X	X		X	X		4
Performance requirements for managers and supervisors are established through the business planning process; reviews are conducted quarterly	X						1
Strong evidence of being a good neighbour across the entire organization	X	X			X		3
There is a clear focus on aligning all corporate activities to meet these strategic goals		X	X	X			3
Top management provides significant resources for learning and recognition to support and provide employee contributions		X				Need for improvement 5Xs = 5 pts 4Xs = 4 pts 3Xs = 3 pts 2Xs = 2 pts 1Xs = 1 pts	1
A committed core of employees who have been on teams have bought into and support management's quality approach	X		X		X		3

Figure 1   Consolidate information sources for strengths.

most. To become the best, businesses cannot only improve what they do poorly; they must also improve what they do well.

The logic holds that if a business were to improve not only its weaknesses, but also its strengths, it would increase its chances of becoming world class as well as financially sound. The key here is to choose only a few areas to improve, not hundreds. To do this, an organization must prioritize, develop action plans, implement, and review. The PDKA model emphasizes the need to improve strengths as well as weaknesses to become truly competitive and to stay competitive. Criteria must be developed to measure priority, that is, which strength areas will be improved and which areas of opportunity will be improved.

Step A6, Prioritize the opportunity areas: Information must be prioritized, because all issues cannot be resolved at one time (Figures 3 and 4). First, the team stratifies this data to eliminate duplication. Once stratification has occurred, the team must determine the priority of these

Consolidate information sources							Example
Feedback summary – opportunity areas							
Comment area	Information sources						
Category 6.0	Federal Quality Institute	Total quality assessment	Site visit	Employees	Employees Internal assessment	Remarks	Problem severity
Limited process measures used	X				X		2
Few trends and current levels of performance are shared	X	X		X	X		4
Environmental measures do not exist other than non-compliances	X						1
Most competitive analysis limited to regional competitors	X	X		X			3
Industry averages used not "best in class" (6.1)		X					1
No benchmark comparison provided related to supplier quality results		X				Need for improvement 5Xs = 5 pts 4Xs = 4 pts 3Xs = 3 pts 2Xs = 2 pts 1Xs = 1 pt	1

Figure 2   Consolidate information sources for weaknesses.

strength and opportunity areas. To assist in the prioritization of all opportunities, the team develops criteria by which to measure priority by using a tool called a problem-selection matrix. The problem-selection matrix has six criteria to address: Impact on customer, Impact on business bottom line, Problem severity (this is straight from the matrices in step A5), Team's ability to solve, Team's ability to solve quickly, and Team's ability to solve with low cost.

For each opportunity area these six criteria are scored on a scale from one (low correlation) to five (high correlation). The scores are totaled and priorities are now determined by number. The highest point getters are worked on. In other words, an organization should work on the few areas of opportunities which will improve the business the most and work on the one or two strengths which, if made better, would put it into a world class category.

Step A7, Select and proceed with opportunity areas: The action plan team must now select the opportunity areas and strength areas to be implemented (Figures 5 and 6). The scale has a possible 30 points which can be totaled for a potential improvement area. The problem-selection scale has been linked to the Baldrige scale, and the cutoff points have been set at 30% and 70% on the Baldrige scale. What this means is that areas of opportunity whose scores fall below 30% (21 points) or areas of strength whose scores rise above 70% (21 points) on the Baldrige scale

| | Prioritize strength areas | | | | | | Example |

Problem selection matrix (scale 1,2,3,4,5 (1 = low, 5 = high)							
Strengths	Impact on customer group, F&HP, senior management or F&HP employees	Impact on business bottom line	Need for improvement	Teams' ability to solve	Teams' ability to solve quickly	Teams' ability to solve with low cost	Totals (add total scores together)
Senior managers have researched (benchmarked), developed, and are actively leading the deployment throughout the organization	1	5	5	5	4	4	24
Managers throughout the organization participate in communicating and reinforcing quality values	3	3	4	5	3	3	21
Performance requirements for managers and supervisors are established through the business-planning process; reviews are conducted quarterly	1	5	1	5	4	4	20
Strong evidence of being a good neighbour across the entire organization	1	3	3	5	5	5	22
There is a clear focus on aligning all corporate activities to meet these strategic goals	2	3	3	5	5	5	23
Top management provides significant resources for learning and recognition to support and provide employee contribution	2	4	1	2	2	3	14
A committed core of employees who have been on teams have bought into and support management's quality approach	1	4	3	2	2	3	15

Figure 3   Prioritize strength areas.

| | Prioritize opportunity areas | | | | | | Example |

Problem selection matrix (scale 1,2,3,4,5 (1 = low, 5 = high)							
Problems or opportunities	Impact on customer group, F&HP, senior management or F&HP employees	Impact on business bottom line	Need for improvement (problem severity score comes from list of opportunity areas)	Teams' ability to solve	Teams' ability to solve quickly	Teams' ability to solve with low cost	Totals (add total scores together)
Limited process measures used	1	5	2	5	4	4	24
Few trends and current levels of performance are shared	3	3	4	5	3	3	21
Environmental measures do not exist other than non-compliances	1	5	1	5	4	4	20
Most competitive analysis limited to regional competitors	1	3	3	5	5	5	22
Industry averages used not "best in class" (6.1)	2	3	1	5	5	5	21
No benchmark comparison provided related to supplier quality results	2	4	1	2	2	3	14

Figure 4   Prioritize weakness areas.

Strengths	Totals	
Senior managers have researched (benchmarked), developed, and are actively leading the deployment throughout the organization	24	
Strong evidence of being a good neighbour across the entire organization	23	
There is a clear focus on aligning all corporate activities to meet these strategic goals	22	
Performance requirements for managers and supervisors are established through the business-planning process; reviews are conducted quarterly	21	
Managers throughout the organization participate in communicating and reinforcing quality values	20	
A committed core of employees who have been on teams have bought into and support management's quality approach	15	
Top management provides significant resources for learning and recognition to support and provide employee contribution	14	

Select strength areas

Example

Stratify problem selection matrix

Total scores:
21 pts or greater: do now without question

Figure 5 Select strength areas.

Problems or opportunities	Totals	
Most competitive analysis limited to regional competitors	22	
Limited process measures used	21	
Few trends and current levels of performance are shared	21	
Industry averages used not "best in class" (6.1)	20	
Environmental measures do not exist other than non-compliances	20	
No benchmark comparison provided related to supplier quality results (6.4)	14	

Select opportunity areas

Example

Stratify problem selection matrix

Total scores:
21 pts or greater: do now without question

Figure 6 Select weakness areas.

are considered critical. Thus, every opportunity area and every strength area that scores equal to or higher than 21 points is attacked without question. Opportunity areas and strength areas that fall below a score of 21 points may or may not be pursued at this point.

At this point in the process, the information from feedback sources has been integrated with theory (in this case the linking of the problem-selection matrix with the Baldrige categories) to create knowledge. In other words, Knowledge can now be used to Act.

Step A8, Schedule and budget the selected opportunity areas: Next the action plan team uses the list of prioritized improvement opportunities to determine how many feasibly can be done within given time and dollar constraints. The team schedules and budgets the selected opportunity areas and performs a cost/benefit analysis for management. The opportunity areas all have a projected scheduled completion and are tracked with actuals. Presentation and review dates are also noted on the schedule.

The team also estimates potential costs to implement a solution, so that when it is time to implement, money has been budgeted. Once management reviews the proposed actions and the cost/benefit analyses, these tasks are factored into each organization's business plan actions, and tasks are assigned to deploy improvements.

Step A9, Assign selected opportunity areas as tasks: The action plan team now assigns all selected opportunity areas as tasks which can now be factored back into the business by using the six-step problem solving process or a more appropriate method to implement the opportunity. Some tasks can be improved another way: for example, a task may already be being worked on by a team; or an organization may be running a pilot on a problem; or a problem can be resolved by an individual or work unit. In these cases, the effort does not need to be duplicated or the problem solved with the six-step problem-solving process. However, the existing efforts need to be tracked in the organization's business plans. For those opportunity areas which will be assigned as tasks, the action plan team now must select a sponsor, team leader, team members, and review team members.

The sponsor is a manager who has ownership of the problem and who will take on the responsibility for helping the team succeed. The team will use the six-step problem solving method to drive the problem to root cause.

Step D2, Form Task Teams?: The action plan team decides whether to form task teams or not.

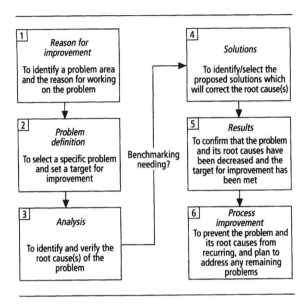

Figure 7   Six-step, problem-solving process.

Step A10, Implement improvements using a problem-solving process: Task teams are improvement teams whose members are appointed by management based on their expertise and are assigned a problem to work on. Each task team uses the problem-solving process (Figure 7) to attack the root cause for each area of opportunity. Deployment teams are sets of task teams that are assigned to work on the selected improvement opportunities of greater complexity than those handled by a single task team. Teams disband when the results are satisfactory and the process has been standardized.

Step A11, Use another method to implement opportunities: The action plan team has determined that task teams are not the appropriate way to implement this improvement. The team now chooses a more appropriate method for implementing the improvement area, such as conducting a pilot, borrowing a world-class process from another company, or having an individual or work unit come up with a solution.

Step A12, Integrate selected strengths and opportunities into business plans: Once strengths and opportunities have been integrated into business plans and their progress tracked, the PDKA cycle has been successfully completed.

Step A13, Conduct reviews: Management conducts Business Reviews, which include the status of the task teams and the implementation of the solution for the opportunity areas.

CONCLUSION

Results: Correlation to the Bottom Line

Becoming a world-class company (650–1000 points on the Baldrige scale) requires converting information into knowledge. A company can implement solutions based on knowledge by using the PDKA process. It is critical that a company learn to complete the cycle and to implement solutions to problems, because it moves a business toward world class, which equates to improving customer satisfaction and improving the bottom line.

If we want to improve the bottom line of our businesses, we cannot simply assess our businesses and hope for the best. We must convert our information to knowledge; then we must prioritize and implement our findings. No longer can we complete only steps 1 (plan to assess our business systems) and 2 (do an assessment of our business systems and gain feedback) of the PDKA method, we must convert our collected information into knowledge. We must complete step 3 (converting information to knowledge) and step 4 (acting on our newly obtained knowledge by deploying solutions and tracking results). By using this four-step PDKA model, F&HP company has improved its bottom line—an average net savings per year for the last 4 years of $30M. The company has also won the Tennessee Quality Award; has been awarded Finalist for The President of the United States Quality Award; has had multiple teams selected as semi-finalists and one team awarded as finalist for the Rochester Institute of Technology/USA TODAY Quality Cup; and one team has won the RIT USA Today Quality Cup. F&HP has and also been awarded the Hammer Award for leading the way in Reinventing Government. The results speak for themselves: Converting information to knowledge, and using knowledge to drive improvements pays.

This case study shows how feedback from Baldrige-style assessments, along with Information from other sources can be integrated with theory (Baldrige and problem-selection matrixes) to create Knowledge (step 3 of the PDKA Model). This Knowledge can then be prioritized and implemented through the business planning process to Act by deploying solutions and track results (step 4 of the PDKA model). This case study also shows that companies that develop their strengths as well as improve upon their weaknesses are more likely to become world-class than those who improve only their weaknesses. The Presidential Award Finalist upon whom this case study was based, feels that other companies can benefit from implementing the PDKA model.

COMPANY

National Aeronautics and Space Administration (NASA), Kennedy Space Center

PRODUCTS AND SERVICES:

- Preparation, test and checkout of launch systems and payloads such as satellites and scientific and technological experiments
- Coordinate countdown and mission safety in launch, on-orbit activities, and landing and recovery operations

LOCATION:

Cape Kennedy, Florida

NUMBER OF EMPLOYEES:

1,900 NASA employees
16,100 contract employees

KEY PERSONNEL:

Case Study Contributor: Donna Cox

KEY USES OF INFORMATION:

- Internal self-assessments use President's Award criteria
- External assessments
- "Town meetings," surveys, and one-on-one interviews with employees
- Florida's Sterling Award feedback
- Benchmarking

ORGANIZING FRAMEWORK TO CREATE KNOWLEDGE:

KSC Information and Analysis System

AWARDS AND SIGNIFICANT RESULTS:

PQA Improvement Prototype Award

Winning the President's Quality Improvement Prototype Award was a significant recognition of our focus on quality, adding value for our customers and making our center's civil service and contractor work force the best in the world. In 1994, KSC was a finalist for the award. Continuing to make significant improvements, the nation's spaceport was recognized as a winner in 1995, and named a finalist this year for the Presidential Award for Quality.
— Gene Thomas
Deputy Director
Kennedy Space Center, 1996

INTRODUCTION AND BUSINESS OVERVIEW:

The National Aeronautics and Space Administration (NASA) is in the midst of a process of evolution and change, largely dictated by tight budgets, dwindling resources and a shrinking work force. Change has been a hallmark of KSC from the first manned launches of Mercury, Gemini, and Apollo and into the Shuttle program of today. A more drastic change now exists: one from NASA having the lead role in the launch to one of contractors assuming more of the responsibility through privatization, which will result in a significant reduction of the civil service work force. Adapting to change while continually making improvements is the way business is done at the John F. Kennedy Space Center.

President Clinton and Vice President Gore have pointed to NASA as a prime example of their National Performance Review initiative to "reinvent government" by reducing bureaucracy, cutting costs and improving efficiency. In line with NASA Administrator Daniel Goldin's pledge to fulfill America's space and aeronautics mission in "a faster, better, cheaper way," Kennedy Space Center has been at the forefront of NASA's reinvention process. Vice President Gore recognized KSC, in particular, at a 1995 celebration of the NASA Heroes of Reinvention. The Vice President said, "the Shuttle and payload customer support team have really been making things work better and cost less in the last few years, the Shuttle team, including a terrific group of contractors led by Lockheed, has cut the cost of each flight. . . . They get the Shuttle assembled and on the launch pad 40 percent faster with one-third as much labor. This all adds up to a savings of $340 million per year. And quality and safety are way up at the same time."

Quality management has evolved as an integral part of our operation since KSC was established more than 3 decades ago. Senior managers set the pattern of personal commitment to KSC as a customer-focused organization devoted to quality management and Continual Improvement. Emphasis was placed on six areas:

- goal setting
- planning
- reviewing quality and operational performance
- communicating with employees and encouraging involvement
- recognizing employee contributions
- customer focus

A major impetus for change at KSC grew out of the complex task of trying to cut costs and maintain an ambitious launch schedule, while emphasizing safety as the top priority, when the Space Shuttle returned to flight status 32 months after the 1986 Challenger accident.

Former KSC Center Director Robert L. Crippen in 1994 recalled that "the need to improve the process took on added urgency in 1989 when returning to flight status after Challenger, our flight capability was only three per year. Flight rate projections were ramping up to 12 a year. The reaction to Challenger was to go to an overly conservative approach to preparing for launches. Shuttle processing was mired in excessive paperwork and redundancies. We resolved to correct these excesses while maintaining safety as our number one priority and turned to Total Quality Management/Continual Improvement (TQM/CI) for a solution."

Armed with a Strategic Plan, a well-defined mission and vision, and a Continual Improvement process road map, KSC has made considerable strides in process improvement, particularly in operational efficiencies to cut costs, by improving labor utilization and reducing cycle times. In 5 years, we reduced the cost per Shuttle flight by $164 million, or more than 50%. During the same period, we reduced the total number of labor hours per Shuttle mission by 800,000, a reduction of 50%. These operational efficiencies were achieved through the efforts of hundreds of Continual Improvement teams that baselined the process, using TQM tools and techniques to develop improvements, reviewing customer feedback, and making adjustments when necessary.

"Launch work is teamwork!" is more than a motto at KSC. KSC recognizes that teamwork operating in a culture of continual improvement is the only way to successfully accomplish its mission and manage the complex, interdependent processes and multiple facilities required to process and launch the Space Shuttle and its payloads.

After more than a decade of Shuttle missions, human space flight remains an extremely risky enterprise with literally thousands of components that must operate in perfect unison during launch and orbit. The Space Shuttle is admittedly the most complex machine ever built. The astronauts corps, one of KSC's internal customers, face this risk and trust the KSC team with their lives to remain and launch this incredible space vehicle on each mission.

Similarly, the processes for "preparing, launching and landing each mission" are among the most complex and risky in the world. KSC's team deals with this complexity and risk on a daily basis. Processing requires the scheduling of over 1,500 tasks with nearly 20,000 constraints that must be satisfied before the work is complete. When three Orbiters are processed simultaneously, which is usually the case, over 4,500 tasks and 60,000 constraints are orchestrated and frequently adjusted due to the discovery of unplanned work. In addition, many of the tasks are hazardous operations, involving toxic liquids and gases and explosives.

All of the work involved in ensuring a successful Space Shuttle mission comes together at the Kennedy Space Center. That will continue to be true for NASA's next biggest program challenge—America's participation in the international Space Station program. The KSC journey to excellence evolved from a tradition of successful teamwork built over more than 3 decades. It is a journey of discovery that KSC hopes will never end.

MISSION

The John F. Kennedy Space Center is a world leader in space launch, landing, and payload processing operations. It is the home to NASA's

Space Shuttle fleet of four orbiters—Atlantis, Columbia, Discovery, and Endeavor—which transport astronauts and a wide variety of payloads into earth orbit and beyond, and prepare for their return. KSC also provides launch support and oversight for activities at adjacent Cape Canaveral Air Station in Florida, at Vandenberg Air Force Base in California, and at contingency and secondary Space Shuttle landing sites all over the world. Kennedy Space Center is the nation's only processing and launch site for its manned space vehicles, and it is the final preparation and integration site for payloads, as well as the prime landing site. KSC is America's gateway to the universe.

The integrated KSC team of NASA and contractor employees is responsible for the processing, testing, and launching of the Space Shuttle, and the landing and recovery of the Space Shuttle orbiter and payloads, as well as the recovery of the reusable solid rocket boosters. By the end of 1995, KSC served as the launch site of 73 Space Shuttle flights, the landing site for 27 missions, and processing site for hundreds of payloads since the Shuttle first flew in 1981.

KSC is gearing up for its next major task as the final processing and integration point, as well as launch site, of elements of the international Space Station.

Another important responsibility of the space center is managing an annual budget in excess of $1.3 billion, and an integrated team of slightly more than 14,000 civil service and contractor employees.

Employees/Team

KSC and NASA have always considered all of their employees as their greatest asset. Since the early days of the space program, the work force has exhibited an unparalleled pride in their work and a spirit of cooperation which makes the KSC group the most efficient and effective space launch team in the world.

KSC employs a culturally diverse and skilled work force, provides a work environment that encourages career and personal growth, and rewards and recognizes employees for quality performance. As of January 1996, KSC had 2,200 civil service employees and approximately 12,000 contractor employees representing nearly 60 companies, many of them small and/or minority-owned businesses. Our culturally diverse work force represents all minority groups and encompasses the broadest spectrum of job classifications. More than 75% of the NASA employees, and a significant number of contractor employees, are engineers, scientists, or in other professional disciplines. Continual Improvement (CI) principles and practices have been incorporated throughout the space center and recognized as the best and only way to do business at KSC.

Products and Services

KSC's major products and services include the preparation, test, and check-out of launch systems and payloads such as satellites and scientific and technological experiments. This includes coordinating a well-planned countdown to assure mission safety and success in launch, on-orbit activities, and landing and recovery operations. KSC also is responsible for turnaround operations. Space hardware returns for evaluation, servicing, and repair and refurbishment, if necessary, to be readied for another flight.

Facility and ground support equipment design, construction, maintenance, and logistics are key support services, along with research and development activities for science, engineering, and technology applications. Our personnel are dedicated to enhancing the safety, reliability, efficiency, effectiveness, and quality of KSC services and activities in our CI culture.

Products and services provided for our internal customers include planning, designing, operating, and maintaining the center's power and lighting system, water and sewage systems, waste disposal, heating, air conditioning, communications, buildings and structures, libraries, office automation, fire, security, law enforcement, aircraft and railroad operations, transportation, roads and grounds, food services, mail, child care center, credit union, medical, and other services required for a small city.

Facilities

Kennedy Space Center is located on 140,000 acres surrounded by the Merrit Island National Wildlife Refuge. Additional facilities used by NASA are located on the adjacent 15,804-acre Cape Canaveral Air Station. Resources committed to accomplishing KSC's goals include capital equipment valued at $876 million, 882 facilities with a combined value of $3.2 billion to process payloads, launch and land Space Shuttles; and $2.8 million invested in training employees to maintain its capability to meet the needs of current and future high-technology missions.

Customers

KSC has a diverse group of customers and stakeholders in private industry, academia, and the public sector. Its principal external customers are the international science community; the astronauts who fly on the Shuttle; other NASA centers whose hardware and software become a part of our mission; the media; Congress and the Executive Branch, local, state, and federal regulatory agencies; industry and academia; and ultimately the

American taxpayer. With the international Space Station, NASA and the U.S. have committed to several foreign governments and space agencies as longtime space partners and customers.

The principal customers for KSC, identified for the National Performance Review Reinvention Lab, include direct Shuttle and payload processing customers and the general public. KSC's direct payload customers include universities, industry, other NASA centers, other government agencies, and international customers. Its general public customers include the local, national, and international media; launch and landing guests; educational institutions; teachers and students; and Spaceport USA visitors. Since KSC has the infrastructure and services of a small city, every employee and visitor to the space center also is a customer.

INITIAL IMPROVEMENT EFFORTS

While recognizing the need to improve quality and procedures, yet simultaneously cut costs, KSC senior managers have never veered from the course of placing safety first; there is no higher priority. Initial improvement activities focused on four areas: reducing the number of Shuttle and payload processing activities; strengthening the partnerships with our customers and suppliers; expanding employee involvement/empowerment; and using Total Quality Management (TQM) principles and tools to accomplish these changes.

KSC's first formal centerwide Strategic Plan, developed in 1987, addressed the future of the space center as an integral part of our nation's pursuit of civil space initiatives. During this period, the year after the accident which resulted in the loss of the orbiter Challenger and its crew, KSC focused on the safe return of the Space Shuttle to flight status, and on strengthening the management and technical teamwork necessary for fulfilling its mission.

Quality improvement efforts accelerated significantly in 1990 to improve processes and eliminate redundancies and excessive paperwork. TQM became a key element to be used to successfully accomplish its mission. KSC-sponsored briefings were given by distinguished and notable TQM advocates such as Philip Crosby of Philip Crosby Associates, Tina Sung of the Federal Quality Institute, and Malcolm Baldrige National Quality Award winners David Kearns of Xerox, and Chris Holloway of Milliken.

In 1991, KSC contracted with The Cumberland Group, a national consulting firm experienced in working with the government and the aerospace industry. Its purpose was to identify and develop a base for implementation of TQM, as well as providing extensive employee and management orientation and training. Workshops were conducted throughout the organization for employees at all levels. That year, the space center also participated in a NASA internal self-assessment and evaluation using the President's award criteria.

By the end of 1991, most employees had attended the workshops and received TQM training. These workshops included developing mission, vision, and value statements and overall TQM implementation planning. KSC has continued to present workshops appropriate for new employees and those who were previously unable to attend the initial presentations.

The Strategic Plan rewritten in 1991 included more specific goals and an added emphasis on TQM as the way to do business. A management instruction was written and later revised to implement "an approach to managing work using the entire work force in the analytical evaluation of all work processes for the purpose of making continuous improvements in quality and productivity." Recognizing that quality management is an ongoing process of improvement, KSC at that time elected to use the term "Continual Improvement" as synonymous with Total Quality Management.

In 1993, a NASA team developed a new mission statement with input and support from our employees, customers, and suppliers. Since 1993, representatives appointed from each functional area have convened annually to review and recommend revisions to the KSC Strategic Plan in accordance with its mission statement, vision, and goals. Another team was formed to develop its Continual Improvement plan to support the KSC mission, vision, goals, and objectives contained in the plans of all KSC line organizations.

The 1995 plan drew from five basic sources to identify potential activities, impacts, or changes. First, documents such as the NASA Agency and Office of Space Flight Strategic Plans were analyzed as they drive requirements down to the Center level. Second interviews were conducted with first-line directors using specific questions to determine potential activities, impacts, or changes for our future. Third, KSC conducted interviews with the chief executives of its three major contractors. Fourth, the KSC government union was interviewed with similar questions. Fifth, a periodical search, encompassing articles written over past months, was conducted for external trends and indicators.

From this effort, approximately 75 areas were identified for review to ensure adequate coverage in the KSC plan and for consistency with the agency documents. The plan encompasses short- and long-range goals. The plan considers (1) customer/supplier requirements; (2) work process changes from our continual improvement efforts; and (3) our unique role in this nation's civilian space effort.

OPERATIONAL IMPROVEMENTS

Kennedy Space Center has excelled in process improvement, employing TQM techniques and teamwork to get results. Deputy Director Gene Thomas pointed out: "The first integrated NASA/contractor team within NASA was chartered at KSC in 1984 to reduce corrosion damage to our huge structures and outside equipment resulting from the harsh ocean environment. During the past 10 years, the team has saved around $5 million and has developed techniques and products which are used throughout the world."

KSC recognized that quality management was an integral part of the TQM process and that its objective should be to improve its processes and not just fix problems. Improvements were continually sought in all operations and activities. "We need to constantly reexamine how we do things and continue to find new and even better ways to do them," said Center Director Jay Honeycutt, who personally oversaw major improvements in Shuttle processing as the former Director of Shuttle Operations at KSC.

One of the NASA engineers involved in those improvements was Eric Redding, whose CI team reduced the time needed to process solid rocket boosters (SRBs) in the Vehicle Assembly Building from more than 70 days in 1990 to an average of 20 days, or as few as 18, now. Yet there is still room for improvement.

"Even though we've realized some great efficiencies in SRB and external tank processing, the team is not going to stop and rest on those laurels. They are continuing to work on enhancements and better ways to do things, especially in the upcoming years and with more stringent restrictions, additional budget cuts and fewer resources. Team members are continuing to explore ways to improve the records they've already set," Redding said.

As a part of its planning, KSC has taken the lead with the formation of a labor-management partnership council with the government union. The action was initiated in 1994 with training provided by the Federal Mediation and Conciliation Service and a Partnership Council Charter was approved. This was communicated to all civil services employees at KSC. The Council meets monthly to discuss and resolve management and employee issues. Representatives of the Council were invited to speak at the National Partnership Council's meeting in Atlanta, Georgia, in June 1995.

EXTERNAL ASSESSMENTS (KEY INFORMATION)

Center Director Robert Crippen in 1993 wanted an external assessment of KSC's progress in implementing Continual Improvement processes. KSC obtained that external assessment by applying for the Quality Improvement Prototype Award, and was named an award finalist by the Federal Quality Institute (FQI) in 1994.

In that first application, the judges found that KSC had strengths in most areas, and said KSC has "a strong vision of its role, pride in itself, and in the quality of its people and its work. It has a strong customer focus with a variety of good ways for gathering customer feedback data and using those data in ways that relate to the unique needs of each directorate and mission. There is also very clear evidence that employees feel empowered to improve their work processes and satisfy their customers."

The judges listed some reasons for deciding KSC was not ready to be designated a Quality Improvement Prototype winner. Two of those factors were:

■ At that time strategic planning had largely been a top-down effort with limited use of employee, supplier, and customer feedback in our strategic and Continual Improvement Planning.

■ In the area of quality assurance, there was yet to be established a systematic use of analytical tools to assess, control, and improve processes centerwide, nor was there evidence of a systematic approach to assure or assess the quality of externally provided goods and services. The judges wanted to see trend data on external customer satisfaction, data tracking the performance of suppliers, and trend data on employees receiving quality-related training and on employees involved in CI activities.

Each assessment had great value in itself. The areas for improvement identified in each assessment were given a close review, and steps were initiated to strengthen these weaknesses. The assessments also forced KSC's line organizations to take a critical look at their quality efforts at all levels; it educated managers and employees about the meaning and importance of quality; and it provided senior management with an expert impartial, outside "diagnosis" of both our strengths and opportunities for improvement. KSC implemented changes and made sufficient progress in our Continual Improvement efforts to earn a Quality Improvement Prototype Award in 1995.

A summary of our most recent assessment revealed that KSC is strong in all seven categories of the application and that the knowledge of detailed systems implementation of plans and policies including the efficiency of operations, exceeded the documentation presented in its application. KSC's leadership team is highly regarded, all levels of the organization have systems in place to assure that information collected is reliable, consistent, valid, and readily assessable in response to user needs. KSC's strategic plan has existed for many years and identified a number of key strategic improvements with steps taken that have resulted in the attainment of a number of key strategic improvements. The site visit also revealed that customer satisfaction is what every member of KSC is all about, that KSC have a comprehensive, well-documented system for determining current, near-term, and future customer requirements and expectations.

IMPROVEMENT METHODS

I was reassured that it's possible to change the culture at NASA from within. There was no pressure to conform to the existing fundamental structure. We weren't placed under any time pressure and each member was able to present an option they felt should be considered. My input was valued and we all had an equal vote.
— Todd Arnold
KSC CI Team Member, 1993

As the premier site for launch and recovery of Space Shuttles and their payloads, KSC's measurable achievements are visible to the entire world. All employees understand that success depends on "doing the job right the first time." KSC cannot pass its work on to the next location because the work leaves a propulsive mode on its way to outer space. KSC has to do its best because it is the last stop. Management of process quality is considered to be the job of every employee at the space center. CI has become a permanent and integral part of KSC's work culture.

The center director sets the standard for leadership performance through his personal commitment to the highest ethical code of conduct. The principles of leadership at KSC revolve around pride in the U.S., the importance of the KSC mission, adherence to strict safety procedures, and respect for the individual employee. Every employee is considered an essential part of the "KSC team." The success of the team in meeting its ultimate goal of fulfilling Kennedy Space Center's mission requires the close cooperation of all of its members.

Setting the Stage with Strategic Planning and Gaining Key Information

For several years following the Challenger accident, KSC efforts in goal setting and strategic planning helped to focus the space center on its mission and generated many improvements to processes, especially after a formal structure for Continual Improvement activities was initiated in January 1991. Center Director Robert L. Crippen then took KSC a step further.

Declaring that KSC needs to place greater emphasis on the way it works, Crippen kicked off a comprehensive centerwide strategic planning effort to develop plans with all the tools necessary to manage quality and operational performance improvement. He empowered a group of employees representing each directorate to build a new vision, prepare a mission statement including major operational elements, develop updated goals for the next five years, and identify short-term objectives that correspond to the current NASA Agency and Office of Space Flight Strategic Plans. KSC was the first NASA center to have its employees assist in the development of a vision statement, and serve as key contributors to the latest agencywide mission and vision statement.

Developed through "town meetings," surveys and one-on-one interviews with employees and managers at all levels, the employee-owned KSC Strategic Plan is KSC's blueprint for the future. KSC also recognizes that Continual Improvement is an integral part of the Strategic Plan and is necessary to accomplish KSC's goals. A separate Continual Improvement Plan developed by a centerwide team of employees is the road map to follow in making process and quality improvements. KSC's goal was to merge the Strategic and Continual Improvement Plans into one document during calendar year 1996.

Measuring our Progress

The KSC Strategic Plan has been the major driver in the development of KSC's key measurements. In 1994, the mission, goals and objectives outlined in the Strategic Plan were translated into performance objectives for each organization, with metrics established to measure success. Each directorate developed its own supplemental strategic plan related to the overall strategy. These strategic plans are linked to the performance plans for each director. KSC's vision, quality values and customer focus also are translated into requirements in the Individual Development Plan for each employee.

The key types of data KSC uses include quality, efficiency, timeliness, customer satisfaction, safety, environmental, employee development, employee involvement, diversity, and supplier performance.

INFORMATION AND ANALYSIS

Effective information and analysis systems are required to enable KSC to meet the challenges described in the organizational overview with the highest levels of quality, cost, and schedule performance ever achieved at KSC. The information and analysis systems also enable outstanding successes in continual improvement, as reflected in the positive results. This section briefly describes how KSC manages data, turns data into information through analysis, gains additional information on best practices through competitive comparisons and benchmarking, and makes decisions based on this information to get the results required by its customers.

Management of Information and Data

This section describes how KSC selects and manages data used for planning, evaluating, and improving overall performance with respect to customer needs and expectations.

The KSC Strategic Plan guides the selection and development of KSC's key processes and their respective performance measures. The Strategic Plan divides the KSC mission into eight primary processes (tasks) and identifies strategic goals. The strategic goals determine the key business drivers and key processes that allow KSC to meet or exceed customer expectations.

To measure progress toward these goals, performance measures for each key process are established. The performance measures determine the data required to manage and continually improve key process performance. Performance measures are realigned and revised to reflect customer feedback, successfully implemented process improvements, and changes in the strategic goals.

The performance measures for each key process are divided into four main categories: customer satisfaction, quality (including safety), cycle time, and cost. Other measures cover key business drivers such as human resources, community service, environmental protection, technology transfer, cultural diversity, and public affairs.

Several criteria determine the data used for improving performance. For example, the data must (1) be effective in measuring desired performance and in forecasting results; (2) provide information to identify problems and corrective actions; and (3) be economical to collect. A KSC Measurement Handbook and Workbook was developed to assist process owners in planning and establishing their performance measurement programs.

KSC selects data that, after appropriate analyses, supports decision-makers with information directly relevant to the KSC strategic goals.

The overall approach for ensuring reliability, consistency, validity, and ready access of data is increased use of automation, computer networks, and common computer databases.

For example, KSC employees use electronic mail to communicate and to send/receive data files. Several data systems give users the flexibility to customize analyses and online reporting of data they retrieve. Work and administrative procedures may also be accessed online. Automated tracking systems provide timely data on work location, in-work status, and estimated completion dates. User requirements are also driving the expansion of KSC Internet capabilities.

Many different methods are used to provide data with integrity in information and analysis systems. Data integrity includes data accuracy, consistency, reliability, validity, and completeness. All data systems employ appropriate electronic security measures. Work documents, part inventory tags, and warehouse locations are bar-coded to enhance traceability, reduce data entry errors, and improve data timeliness. Several databases employ automatic data entry edit checking.

Classes are available for training users to properly fill out data entry forms and to use new computer systems. A computer-based tutorial explaining the importance of data integrity and how shop floor data fits into the continuous improvement cycle is available to Shuttle processing employees in their work areas over a local area network.

KSC established an early standard for effective information and analysis systems with the Launch Processing System, which has supported our most critical process—flight hardware test/checkout and launch operations—since the Apollo era. The Launch Processing System routinely collects enormous amounts of reliable data, analyzes that data and makes it rapidly available to users. This system was designed to meet the extraordinary demands of its users, the launch team. The tremendous launch success KSC has achieved over its history is, in part, attributable to this extremely effective information and analysis system. The basic design approach of the Launch Processing System was adopted by KSC managers as information and analysis systems in other areas were refined through the use of computer systems.

All KSC organizations develop their own key quality and performance indicators to allow them to identify areas needing improvement. KSC managers use various methods to align and integrate data analyses with the key business drivers. The award fee criteria and areas of emphasis help identify specific key business drivers for each six-month period, allowing managers to focus their improvement efforts. NASA and contractor functional reviews allow management to monitor progress and make necessary adjustments to achieve program success.

KSC evaluates and improves methods for the selection of data through its analysis results and through feedback from customers, users, and process improvement teams. For example, the payload customer survey questions are updated annually based on customer feedback. The award fee evaluation report provides valuable feedback from NASA to its contractors. This report details contractor strengths and areas for improvement. Process improvement teams are chartered by management to determine root causes of substandard process performance and to make recommendations for corrective actions.

Competitive Comparisons and Benchmarking

KSC comparative studies cover the full range of the benchmarking spectrum. In general, informal and/or internal benchmarking studies, which have been a "way of life" at KSC for many years, are setting the stage for more formal external benchmarking efforts. KSC is also developing partnerships and innovative techniques to make formal external benchmarking efforts more cost-effective for KSC. It will, however, continue to perform different types of comparative studies to meet varying user needs. This section describes the KSC efforts to build a sustainable process for benchmarking and competitive comparisons of key processes.

Competitive comparisons and benchmarking studies are performed in most functional areas of KSC. An external study refers to a study with partners outside KSC, while an internal study refers to a comparison of processes between various organizations or facilities within KSC. It should be noted that several of KSC's internal studies would qualify as external studies in organizations less diverse than KSC. For example, the benchmarking study for government property management is internal to KSC but involves seven different contractor organizations.

The specific results of each study vary depending on the process studied, the number of process improvements implemented, and how long the improvements have been in place. For example, the complete results of the government property management benchmarking study are not yet available, but early indications are that cost savings will be significant. Within 2 months of the distribution of findings to process owners, three organizations reported a combined cost avoidance of over $41,000. A fourth organization reported a 57% reduction in cycle time for Property Loss, Damaged, or Destroyed (PLDD) reports, and a fifth organization reduced the number of PLDD reports by 84%.

The key processes and performance measures identified through the KSC strategic planning process also form the foundation of our comparative study efforts. In general, the amount of resources invested in a benchmarking study is directly proportional to the level of potential payback. Additional criteria for selection of processes for benchmarking studies include process criticality, process stability, availability of documentation, and cost of implementing process changes.

Internal benchmarks are regularly identified within KSC's diverse organizations and facilities by analyzing and comparing the internal performance measurements. For example, the number of technician work-hours per processing flow are routinely compared between the three orbiter processing facilities.

In addition, process owners are encouraged to constantly be aware of developments outside KSC through informal networks, technical and business literature, and performance evaluation by internal NASA/government audits and external independent organizations, such as the Software Engineering Institute for the assessment of software development processes.

KSC continues to refine its benchmarking methodologies based on customer feedback and requirements, the availability of new techniques, and evaluation of improvements and lessons learned from previous benchmarking studies. The result is several innovative benchmarking efforts that place KSC on NASA's "cutting edge" of comparative study techniques.

For example, the NASA Facilities Maintenance Benchmarking Group meets monthly, via the NASA video conferencing system, with representatives from NASA centers and industry. During the video conferences, the team's approaches are constantly updated. The video conferencing system is a KSC institutional capability, so this benchmarking effort itself contributes to our strategic goal to "better utilize KSC's institutional capabilities."

The KSC Benchmarking Network, a collaboration of NASA and nine major KSC contractor organizations, has pioneered and customized consortium benchmarking techniques to enable more cost-effective benchmarking studies. The methodology of the KSC Benchmarking Network was tested with a pathfinder study of government property management which has already resulted in the cost savings.

The efforts of the KSC Benchmarking Network Team have been recognized both inside and outside NASA. At the 1995 NASA Continual Improvement and Reinvention Conference, NASA Administrator Dan Goldin presented an award recognizing the team's "outstanding contribution to NASA quality management and the continual improvement philosophy." The Network is also a recipient of a prestigious 1995 Benchmarking Award from the International Benchmarking Clearinghouse.

The Center Director's Discretionary Funds were used to sponsor benchmarking research at a local university. The purpose of the research was to advance the state of the art in benchmarking technology by applying existing benchmarking techniques to Shuttle processing activities and customizing those techniques as required. The results of this research were entered for a 1996 International Benchmarking Clearinghouse award. KSC has undertaken efforts to standardize the structure and improve the effectiveness of formal benchmarking studies. One method is increased benchmarking training from the American Productivity and Quality Center. To date, over 100 employees from all major KSC organizations have been trained. A second method is the adoption of a NASA-wide benchmarking policy to produce more consistent benchmarking results throughout the agency as well as within KSC.

Benchmarking studies are also being refined with external inputs, such as feedback from the 1995 President's Quality Award and Florida's Sterling Award applications. The information from the feedback reports has already been incorporated into the KSC benchmarking strategy.

Perhaps the best indicator of the success and maturity of KSC benchmarking programs is that other organization are now visiting KSC to search for best practices. For example, other NASA centers are studying KSC's benchmarking processes, external industries are studying KSC's safety practices, Texas Instruments is performing a comparative analysis of customer satisfaction with KSC's payload processing organization, and Commonwealth Edison has studied KSC's facility modification/management processes.

Analysis and Use of Organization-Level Data

KSC performs effective analyses of internal and external data to provide quality information to decision-makers at all levels of KSC. The impact of this effective use of organization-level data on KSC's strategic goals is obvious. KSC has continually improved efficiency and customer satisfaction over the past several years without adversely affecting mission success.

The types of analyses performed on organization-level data are driven by user requirements and the types of data collected. Pareto, trend, cause-and-effect correlation, statistical process control, and root-cause analysis techniques are used extensively for organization-level data. Correlation between related performance measures also verifies various analysis results. KSC has the capability to perform more sophisticated analyses when required by the users. Examples of advanced analysis techniques include process simulation modeling (recently used to analyze specific orbiter spares processes), decision modeling, and probabilistic risk assessment.

The organization-level review, action, and planning process includes data and information gained through the internal measurements and the comparative studies described. All data and information is synthesized during several reviews and analyses. For example, quality, schedule, and customer data are aggregated in the manifest/flight planning cycle, the "readiness" reviews, and the payload lessons learned process.

During the manifest/flight planning cycle, payload-unique customer requirements are combined with standard Shuttle requirements and non-standard work to resolve open problems, develop the schedule, and develop work plans. Customer feedback on additional processing time requirements is incorporated. This cycle includes reviews of prior as-planned/as-run schedules to highlight areas needing improvement.

At readiness reviews, quality metrics and overall performance data are presented for NASA management and payload customer review. The readiness reviews are major milestones during the preparation of each Shuttle mission, and they provide an excellent mechanism for KSC to openly communicate with its primary customers.

During the payload lessons learned process, inputs from customer surveys, quality and safety data, and operational performance are reviewed to determine both problem areas and internal best practices.

Organization-level data reviews involve a combination of measures (cost, cycle time, quality, customer satisfaction, and others). All measures are tied to key business drivers and overall customer satisfaction. For example, the readiness reviews emphasize schedule performance and technical issues. Quality inspection trend data is reported at monthly intervals to KSC management. Negative trends result in actions to identify root causes and corrective actions such as procedural changes, design changes, and process changes.

Managers regularly check with customers to assess customer satisfaction and are readily available if the customer has a concern that cannot be satisfied at the working level.

KSC's survey process also provides a measure of customer satisfaction. Administered by an objective group in the organization, each customer payload team is surveyed after each launch. (A launch may include multiple payloads, each with multiple team members.) The survey requests that the customer rate each applicable item in the survey on a scale of 1 (poor) to 5 (excellent). Any score below 4.0 becomes a priority candidate for improvement. The customer can also make narrative comments in the survey.

Because of the nature and uniqueness of payload processing, it is difficult to collect customer satisfaction data that directly compares KSC with other organizations at the overall organizational level. However, KSC regularly collects customer satisfaction data at the process or function level and, to the extent possible, makes comparisons. There are some indirect and anecdotal comparison examples that are relevant. For one group of customers, KSC's major support is the provision of facilities and services.

KSC's metrics indicate that its facilities and services have always been available on schedule and have never delayed a customer's launch, even though some customers have required some extremely stringent standards for cleanliness, temperature, and humidity levels. A customer's mission has never been compromised or a customer's equipment damaged because KSC violated these requirements. There is data from other service providers to compare this performance directly.

A recent analysis by an independent firm sponsored by NASA Headquarters demonstrated that KSC's facility maintenance program was rated significantly higher than the national average and that KSC is among the leaders in implementing new maintenance technologies.

KSC updates and improves its customer survey by analyzing response rates to particular questions, by comments from the customers about their understanding of the survey questions, and by looking for patterns in the narrative responses over several payloads. Survey questions are sometimes expanded to address areas of concern that past customers have identified. As a minimum, the survey instrument is modified annually.

KSC concentrated on the perceived major problem areas during the first few years and achieved a measure of success and improvement. These problem areas were identified because they had "low" scores or because multiple customers made similar comments. The past could be generally characterized as the time when KSC took pride in compliments and fixed problems as fast as possible. KSC is now in the era when it is striving to make excellent performance even better. KSC still takes pride in compliments paid, and it takes every customer comment and complaint as an opportunity for improvement. Each customer comment is addressed individually. When possible, the customer is informed of the proposed improvement, and feedback is solicited. In the past, some comments may not have been resolved for up to a year after launch. KSC's goal now is to resolve all comments within 30 days of receipt.

One key method KSC uses to integrate financial and nonfinancial data is the Associate Administrator Review Status (AARS) process. The AARS process is an organization-level review of data relating KSC's key health indicators (specifically in cost, schedule, and technical performance) to budgetary compliance and customer satisfaction. The AARS review includes discussions pertaining to all completed milestones and communicates any impact on KSC's progress to plan against its designated budget. The AARS also helps determine the budget estimate through fiscal year completion and its associated impact on the rest of NASA.

In addition to the AARS process, quality, operational, safety, socioeconomic, and award fee data are integrated with financial data and reviewed through the Contractor Metrics Report. This report provides timely data to the Comptroller and all levels of KSC management. These reporting processes enable KSC management to quickly resolve financial and non-financial issues and to identify areas needing improvement.

RESULTS OF IMPROVEMENT INITIATIVES

KSC has been on a very successful continual improvement journey, emphasizing strategic planning, implementing measurable improvements, involving employees through teams, and recognizing and involving its customers and suppliers. Its efforts have provided demonstrable results in both internal process improvements and in customer satisfaction.

More Efficient Launch Site Operations

One of the most significant indicators of operation performance is cost per flight reductions. The cost-per-flight demonstrates KSC's improvements in the efficiency and effectiveness of processing and launching Shuttle missions. In just 5 years, the overall cost has been reduced by more than 30% in real year dollars—from $184 million in FY89 to $128 million in FY95. The improvement shown is the result of a broad spectrum of CI initiatives including improved equipment and work package design, better scheduling, and higher production quality.

In order to achieve such a dramatic level of success, these initiatives had to apply to every major step and thousands of procedures involved in preparing the Shuttle and its payloads for spaceflight during the processing flow from the return of the orbiter and arrival of the payloads to checkout at the launch pads to its final countdown and liftoff. We are proud to point out that quality improvements throughout the entire processing flow have resulted in a significant reduction in the amount of time it takes to assemble and integrate a space vehicle for flight.

While improving processing time and reducing costs, KSC has also improved the quality and safety of its delivered flight hardware. This improvement can be documented through the decreasing Shuttle processing incidents found during inspections that are part of KSC's structured surveillance efforts. A significant long-term decrease is demonstrated where the number of processing incidents has steadily decreased from a high of 214 in 1989 to a low of 10 in 1995. When labor hours expended are factored in for a typical processing flow today of 700,000 hours times five flows we find an incredible ratio of 10 incidents per 3.5 million labor hours. This unprecedented performance is a result of heightened awareness, improved training, and better tools and procedures for our work force.

Once troubled with problems and time-consuming delays, stacking problems of the Solid Rocket Boosters have seen a remarkable turnaround as a result of improvement initiatives of several CI teams. KSC's implementation of structured surveillance has optimized the number of inspections conducted by NASA and KSC's contractors. Thousands of inspections and hundreds of checkout operations were conducted by KSC personnel throughout the processing flow to eliminate problems that might occur while the Shuttle is in orbit. KSC has drastically reduced the number of mandatory inspections. This change has been accomplished while maintaining a high quality of delivered flight hardware. The effectiveness of KSC's structured surveillance program has demonstrated a decrease of in-flight anomalies.

Resource Reduction

A major payback of KSC's Continual Improvement efforts is the ability to reduce the number of labor hours required to process a Shuttle and its payloads for a mission while maintaining flight safety. So far, KSC has

- Reduced the total number of labor hours worked at KSC per Shuttle mission from 1.1 million hours in FY89 to approximately 0.7 million by FY95. This is a 36% reduction in average processing time, with an associated total labor cost avoidance of approximately $12 million per mission or roughly $96 million per year that will not have to be paid out.
- Cut overtime in the Orbiter Processing Facility, where most of the Shuttle process work is conducted. Cross-functional process reviews, problem-solving teams, and better planning have led to a 77% reduction in overtime usage from 11% in 1989 to less than 3% in 1995. The direct dollar savings contribution of this effort has been about $20 million per year without any compromise to safety or schedule. In fact, both have improved during this overtime reduction period.

Payload Processing

Payload processing, KSC's second major enterprise, also continues to be more efficient. Efforts focused on preparing Spacelab modules, Spacelab Igloo pallets, and the Tracking Data Relay Satellite for flight have netted a 62%, 42%, and 86% reduction in labor hours, respectively. Similar improvements of 77%, 35%, and 26% in the schedule trends for payload processing have been noted since 1989.

Excess Property Disposal

At the beginning of 1994, KSC's excess property program took over 300 days to dispose of excess property. The standard established by NASA Headquarters was 210 days. The results of the CI team who took on the task to reduce the disposal time to the standard are shown below. The reduction from 300 days to 67 days has helped to showcase KSC as one of the Agency's best and has prompted other Centers to benchmark against KSC's process.

Reduced Energy Consumption

KSC received a Federal Energy and Water Management Award from the Department of Energy (DOE) in November 1994 as a result of conservation efforts during fiscal year 1993. The energy efficiency projects conducted that year led to a savings of $870,000 that did not have to be spent for fuel or electrical power. Efforts during fiscal year 1994 show that KSC is surpassing its Strategic Plan goal of reducing its total energy consumption by 20% by the year 2000. The Kennedy Space Center is also one of the first federal installations to receive DOE funding to retrofit several facilities with improved lighting fixtures and heating, ventilation and air conditioning equipment. KSC continues to exceed its strategic plan goals of energy consumption avoidance. In addition, KSC has claimed rebates from the local power company since November 1993 totaling just over $400,000 for its energy-saving initiatives. Fifty percent of these rebates are available for KSC to reinvest in other energy and water saving projects.

Decreased Environmental Impact

One of the KSC Strategic Plan's major goals is to protect, preserve, and enhance its natural environment. Thirty-two conservation projects have been initiated that incorporate reuse, recycling, product substitution, and recovery methods that have helped the center exceed a self-imposed goal of 30% reduction in hazardous waste. Through the Continual Improvement process, KSC has also developed a chlorofluorocarbon (CFC-113) storage and recycling system that is considered to be one of the best in the world. By employing aqueous cleaning methodologies KSC has reduced consumption of CFC-113 solvent from 648,506 pounds in 1989 to 266,676 pounds in 1994, a reduction of 58.9% and a resultant savings of nearly $1 million.

CONTINUING EFFORTS

KSC is dedicated to the goal of excellence in processing space vehicles and payloads. It has made excellent progress to date, and will continue to press for new and further improvements. The Shuttle process is one of five reinvention activities cited by Vice President Gore's National Performance Review Team in its September 1993 report on improvement efforts in NASA. Vice President Gore praised KSC for its efforts in reducing the costs of Shuttle flights while improving quality and safety.

KSC is using Continual Improvement efforts to reduce the center's cost of operations in response to a shrinking NASA budget, and will continue to do so in the future to meet requirements to downsize and reinvent government. Thanks to its quality improvement efforts, it is able to project that its fiscal year operating costs will be 27% less than in its fiscal year 1991 operating plan.

All KSC employees have been encouraged to actively pursue new and innovative ways to reduce costs, increase efficiency, minimize incidents and waste, and promote productivity improvements in all aspects of our operations. The improvements outlined have resulted in significant cost reductions or resource savings, increased productivity, schedule enhancements, and a better working environment. KSC will continue to improve the way it does its job to assure that its facilities, people, and resources are the most efficient that it can make them.

KSC has learned that while terminology, emphasis, or application of quality culture fundamentals may differ, the quality precepts of successful organizations have much in common.

At this time KSC is facing several new challenges. In the present political environment embracing a strong policy to reduce government employment, balance the federal budget, and privatize to the maximum extent KSC must adapt to a new way of managing its resources and preparing for a new role for civil servants in NASA, while maintaining safety as its number one priority.

In 1995, KSC faced a major challenge of international cooperation in space. The Russian Mir station visit by the Shuttle orbiter was a flawless mission. KSC's teamwork with its Russian counterparts was an example of international customer relationships in total accord. KSC's commitment to teaming as a standard practice worked extremely well as it shared in the successful integration of the Russian hardware into the orbiter.

In the last half of this decade, KSC, with its veteran management team and its exemplary launch and landing processing teams, faces the toughest challenge since the birth of NASA in October 1958. The mandate by the President and Congress to privatize the Shuttle program means that a private aerospace company will take control with NASA taking a "caretaker" role and resulting in a reduction of the civil service work force by one third and a reduction of several thousand contractor employees. KSC's business will be to survey and audit private contractors while maintaining a confidence that the space vehicles and payloads are ready to fly. A literal host of paradigms must change. Safety must not be compromised.

SUMMARY

KSC feels that the commitment of senior management is the single most important factor in the development and implementation of the quality culture at KSC. Over the past several years improvement efforts have continued and matured while the KSC organization and its way of doing business have undergone significant changes due to its senior leadership.

Jay Honeycutt, Center Director, has followed in Crippen's footsteps. As Honeycutt pointed out to the Federal Quality Institute examiners, "The important thing about Continual Improvement is ensuring that it is continual. Just as you (Crippen) carried on as a champion of even greater accomplishments, I hope to add my effort to this momentum."

KSC is confident that it will change as its challenges dictates. The human elements of trust, ethical behavior, and a strong commitment to goals are still prevalent characteristics of KSC employees. The pride in which it does its launch and landing job has not been compromised by temporary setbacks. The employees of KSC are equal to challenges, and their strong record of successes will continue to make the U.S. proud of the accomplishments of the employees of the Kennedy Space Center.

Mindful of the admonition Deputy Center Director Gene Thomas expressed in 1992, "We will continue to consider Kennedy Space Center as a work in progress, adaptive to an ever-changing world and continually on a quest for the very best quality."

"When we stop finding ways to improve how we go about our mission, we will lose our vision."

11

HOW TWO WORLD-CLASS MILITARY SERVICE ORGANIZATIONS IMPLEMENT PRINCIPLES OF PDKA— RED RIVER ARMY DEPOT AND U.S. ARMY ARMAMENT RESEARCH, DEVELOPMENT AND ENGINEERING CENTER, PICATINNY ARSENAL

INTRODUCTION

Both companies that provided these case studies—Red River Army Depot and U.S. Army Armament Research, Development and Engineering Center, Picatinny Arsenal—represent the United States Army. The two organizations vary in the services they provide to the Army. Both of these organizations have embraced the government's initiatives to improve efficiencies. They both show evidence of process changes that have led them to organizational improvements.

Red River Army Depot

Red River is an Army depot that has been actively involved in performance improvement for a number of years. The depot balances three major mission areas: the Ammunition Mission, the Missile Recertification Directorate, and the Maintenance Mission. The depot is a highly responsive and totally integrated logistics power projection platform, capable of rapid and sustained support to America's Army, Navy, Air Force, Marines, and allied forces worldwide. Red River Army Depot stores ammunition; renovates and demolishes conventional munitions; repairs and stores missile systems; receives and ships stock to customers throughout the world; monitors and certifies the readiness of Hawk and Patriot missiles; and rebuilds and refurbishes multi-million dollar Bradley Fighting Vehicles and Multiple Launch Rocket Systems.

Red River Army Depot has also been actively involved in the U.S. government's reinvention initiatives, winning Vice President Gore's Hammer Award and the Army Communities of Excellence Award on numerous occasions. The depot has also won the President's Quality Improvement Prototype Award. In addition to winning awards, Red River has also made some significant improvements that are further described in its case study: it has reduced cycle time, saved $59,000,000 on the Multiple Launch Rocket System Project, and increased production from 3 to 15 missiles per day.

U.S. Army Armament Research, Development and Engineering Center (ARDEC), Picatinny Arsenal

U.S. Army Armament Research, Development and Engineering Center (ARDEC), Picatinny Arsenal provides the U.S. military with the firepower necessary to achieve decisive battlefield victory. It provides America's soldiers, sailors, marines, and aviators with unit and individual weaponry needed to achieve decisive battlefield victory. Its services are delivery of quality weapons, munitions, and fire control systems and associated material.

ARDEC was chosen as a case study because it is a military organization that has taken improvement initiatives seriously. It has been involved in making business improvements for years and has participated in a variety of assessments and programs, one of which was the President of the United States Quality Award Program. ARDEC won the 1996 Presidential Award for Quality, 1996 Quality Partner Award (from New Jersey) and the 1996 Army Communities of Excellence competition, which is now totally based on the Baldrige criteria, and the 1995 Quality Improvement Prototype award. ARDEC was the co-winner of the 1995 Army R&D Organization of the Year, and has been ranked in the top 6 of the 20 Army laboratories in this competition eight times in the last 10 years. ARDEC has also received honorable mention as the Most Improved

Organization in the Army Communities of Excellence Program in 1994. Most recently, ARDEC has gone on to capture the top state award for quality, the New Jersey Quality Achievement Award in 1998 (now renamed the Governor's Award for Performance Excellence). ARDEC also received the CSA, ACOE award in 1999.

HOW THESE COMPANIES HAVE IMPLEMENTED THE PRINCIPLES OF PDKA

Executive Buy-In

Both organizations have recognized the importance of gaining executive buy-in, and both organizations make sure that the senior executives clearly understand the organization's feedback and the need to do something with that feedback.

Red River's leaders are committed to the recommendations and principles of the National Performance Review (NPR). Because of its leaders' strong commitment to continuous quality improvement and a new way of doing business, Red River is recognized as a model for implementing change, training, empowerment, and union management partnerships. The Commander and the top union leader, as well as other top leaders and union officials, have been champions of their quality improvement journey. Red River's union and management moved from the old adversarial way of doing business to the new way of doing business through cooperation, collaboration, and co-ownership. These partnerships provide the support necessary to make change happen. A unique partnership exists at Red River between management and all six of their unions.

ARDEC's leadership is essential to innovation. A strong common purpose and communications among its senior leaders, major customers, and union leaders is fostered through participation on its TQM Executive Council. All of the union leaders are members of the Executive Council, providing insights into the process and partnering on all significant changes.

Consolidation of Information

Red River believes that information empowers, and that you must know where you are before developing a plan for your journey. In other words, you have to be assessed and provided feedback to help you determine what you need to improve. Red River collects data for monthly review and analysis of quality, operational performance, and business plans. There are six major areas of measurement that support the strategic quality and operational business plan:

1. Customer satisfaction/complaint data to continuously identify improvements and current/future requirements.
2. Involvement, since members are their greatest asset.
3. Training to ensure all members are provided empowerment and teamwork skills.
4. Financial data to control revenue and expenses.
5. Production schedules to analyze effectiveness.
6. Performance information to review efficiency, productivity, and waste reduction.

Daily production run charts are kept, and information is then summarized in standardized charts to ensure effective communication and analysis by all levels. Red River has the following key sources of information: self-assessment using Malcolm Baldrige National Quality Award criteria, President's Quality Award Program, alignment of fact with vision, measures for analysis, and data collection of customer satisfaction, member involvement, training cost, schedule, product/service quality, and performance.

ARDEC's key uses of information are summarized as indicators or measurements, benchmarking and competitive comparisons, and the World Wide Web. ARDEC has formal measurement guides and practices in place to monitor metrics based on Malcolm Baldrige National Quality Award and President's Quality Award criteria. Benchmarking the best organizations and processes is a formal part of its measurement process. Its performance measurement process, the Systems Measurement Review, focuses on results, results based on strong quality criteria and information and analysis. ARDEC uses effective feedback systems to obtain knowledge from customers about its products and services. It captures this knowledge through six diverse information sources which are their Listen & Learn strategies: assigning Customer Advocate offices, integrating product teams, matrixing ARDEC employees with the customer, providing formal customer review processes, providing opportunities for informal customer feedback, and obtaining information on similar suppliers.

ARDEC's use of the World Wide Web is considered a Best Practice that provides functions (surveys, feedback, document management, and data in many text and graphic formats) to customers, managers, and other users. Information made available through such technology is of special advantage in business networks and alliances. Feedback of this information assists teams and system owners in improving their measurement processes.

An Organizing Framework

Red River uses a guide developed by the Army Material Command patterned after the President's Quality Award as an organizing framework, while ARDEC has repeatedly applied the Baldrige and President's Quality Award criteria to systematically evaluate and improve its management practices and operational performance.

Clear Decision-Making Criteria

At Red River Army Depot, decisions are based first on fact and then on whether or not they are in line with the vision, quality values, internal/external customer requirements, and what is best for Red River's members.

Assignment of Tasks

Red River forms process action teams to implement improvement tasks. Its Process Action Teams work to improve a specific process and, once the teams complete the task they disband. Over a period of 7 or 8 years, more than 130 Process Action Teams have been formed, with 70% of those being cross-functional.

Problem-Solving Process

Red River has shared values called "HEARTS." HEARTS is an acronym representing traits that identifies its individual and organizational values (Honesty, Ethics, Accountability, Respect, Trust, and Support). Red River's "HEARTS" values are tied to its "HEARTS" training tool. This training tool is a learning process that challenges individuals to go beyond their perceived boundaries, encouraging them to work with others to solve problems. Individuals learn to view obstacles as opportunities. "HEARTS" transcends the organization to become a type of problem-solving method.

ARDEC's entire workforce is involved in problem-solving efforts and contributes to improving work processes. Teams are the backbone of the center's quality and productivity improvement effort. ARDEC makes a significant investment in the selection and training of team facilitators. For many process improvement projects, the team owns the process and is empowered to implement the needed changes. Extensive use of Quality Action Management Boards, Process Action Teams, and Natural Work Groups is one of ARDEC's techniques for channeling employee involvement toward management-selected critical processes to foster a multi-discipline, cross-organizational approach. These teams work toward continuous improvement of ARDEC's major processes.

Progress Tracking

At ARDEC, many of the teams continue to monitor the results of their changes and meet to ensure that the gains are sustained and opportunities for additional improvement are identified and exploited.

Periodic Reviews

ARDEC is considered a Best Practice organization for its strategic planning and effective review process, which simply but most effectively ensures the continual improvement of key metrics within the organization.

CONCLUSIONS

Both case studies discuss senior leadership involvement, which includes not only management but also union representation and participation. Both ARDEC and Red River focus in one way or another on many of the 12 principles of PDKA. ARDEC focuses much attention on leadership involvement and understanding of its business. ARDEC's leadership use indicators to provide key and timely results data. ARDEC also pays keen attention on its Listen & Learn strategies. Red River specifically discusses how it gains senior leader support, how the types of information it gathers and integrates are used, how the organizing framework uses PQAP criteria, how it makes decisions, how it uses process action teams, and how these teams solve problems. Red River believes that it is important to routinely stop, breathe, and refocus efforts. As Red River has looked back at how far it has come, it is clear that its quality journey was and still is like a puzzle coming together. The implementation of just one concept is not enough. Red River focused on its vision and values, and aligned all the pieces of the puzzle to produce the maximum positive impact on its depot's culture and performance. Red River believes that all of the puzzle pieces must be used and integrated as an organization travels its journey.

COMPANY

Red River Army Depot, U.S. Army

PRODUCT/SERVICE:

Store ammunition, renovation and demolition of conventional munitions, repair and storage of missile systems, receipt and shipment of stock to customers throughout the world; monitors and certifies the readiness of Hawk and Patriot missiles. Rebuild and refurbish multi-million dollar Bradley Fighting Vehicles and Multiple Launch Rocket Systems.

LOCATION:

Texarkana, Texas

NUMBER OF EMPLOYEES:

1530

KEY PERSONNEL:

Commander: Col. James C. Dwyer, Jr.

UNION LEADERS:

National Federation of Federal Employees	David Sharp
International Chemical Workers Union	Robert Hopkins
International Guards Union of America	David Hulsey
International Brotherhood of Electrical Workers	Jacky Pate
Federal Firefighters Association	Jewell Smith
United Association of Journeyman and Apprentices of the Plumbing and Pipefitting Industry of the United States and Canada	Bob Tyson
Case Study Point of Contact— Special Advisor to the Commander	Lila Murray

KEY USES OF INFORMATION:

- Self-assessment using Malcolm Baldrige National Quality Award criteria
- Federal Quality Institute's Self-Assessment Guide
- President's Quality Award Program
- Aligning fact with vision
- Measures for analysis (management of process quality, quality and operational results, customer focus and satisfaction)
- Collect data (customer satisfaction, member involvement, training cost, schedule, product/service quality and performance)

ORGANIZING FRAMEWORK TO CREATE KNOWLEDGE:

President of the United States Quality Award Program criteria

AWARDS AND SIGNIFICANT RESULTS:

- Self-managed work teams
- Model for change
- National Performance Review
- President's Quality Improvement Prototype Award Winner (1995)
- Cycle Time Reductions: $59,000,000 savings on Multiple Launch Rocket System Project, $1,500,000 on smoke vehicle prototypes
- Production increase from 3 to 15 missiles per day
- Vice President Gore's Hammer Award
- Army Communities of Excellence Award, winner numerous years since 1991, receiving $1,750,000 in honorarium (Awards in 1991, 1992, 1993, 1994, 1996, and 1998).

INTRODUCTION AND BUSINESS OVERVIEW

The role of the military is changing. Due to major upheavals in the world, the U.S. military has become a smaller, better equipped, more mobile force. The 55-square-mile Red River Defense Complex supports the new military. It accomplishes this by being a highly responsive and totally integrated logistics power projection platform, capable of rapid and sustained support to America's Army, Navy, Air Force, Marines, and allied forces worldwide.

Red River Army Depot is one of the nation's largest defense depots in terms of people and workload. Established in 1941, it originally served only as an ammunition storage area. However, its mission quickly expanded as America's defense planners took advantage of the availability of land, the excellent transportation network, and the depot's close proximity to other important military operations.

Today the depot continues to serve as a vital ammunition storage center and also has an enormous maintenance mission centered around the Army's light-tracked combat vehicle fleet. Activities at the depot are geared to support these mission areas, while providing an outstanding level of installation base support and related services to soldiers and civilian members of its workforce and their families.

Red River Army Depot performs depot level maintenance on a variety of combat vehicles, and weapon and support systems, performs depot-level maintenance, storage, and demilitarization on a variety of ammunition and missiles, and is the sole support to the U.S. Forces and other countries for the recertification of Patriot and Hawk missiles.

Its unique missions include the conversion and modification of light tracked vehicles; rebuild of roadwheel, track, bias and radial tires; design and manufacture of prototype combat vehicles.

Red River's major external customers are its soldiers. Other external customers include the project and commodity managers for the weapons systems mentioned, plus all of the subordinate commands within the Army Materiel Command that oversee its specific industrial activities and processes. Red River's support to these customers includes assistance in design and preparation of technical data packages, parts procurement, fabrication and program management. Red River also works closely with the soldiers in the field, listening to their needs and providing assistance teams to field locations when a personal level of service is needed.

Red River's shared vision is "A Competitive Industrial Complex Excelling in Quality Products and Services." Its two most important strategic goals for attaining its vision are to "Make Total Quality Management Our Management System" and to "Provide Our Customers the Highest Quality Products and Services at the Lowest Possible Cost."

ORGANIZATIONAL OVERVIEW

In addition to a large base operations area, Red River has three major mission areas. In its Ammunition Mission, it stores approximately $6.5 billion of ammunition in a 9,000 acre area. In this secured area, the primary activities are ammunition storage, renovation and demolition of conventional munitions, repair and storage of missile systems, and receipt and shipment of stock to customers throughout the world.

Red River is also the home for the Missile Recertification Directorate, a separate specialized activity that monitors and certifies the readiness of Hawk and Patriot missiles. With its central operations at Red River, this office also operates a field office at Weilerbach, Germany, to further assist worldwide customers. The efficiency and impact of the Missile Recertification Directorate's mission was evident during Operation Desert Storm when the first Scud missile intercept over Israel was accomplished with a Patriot missile bearing Red River's distinctive logo.

Red River's Maintenance Mission is to rebuild and refurbish multi-million dollar Bradley Fighting Vehicles and Multiple Launch Rocket Systems to like-new condition at a fraction of original cost. The vehicles are completely disassembled, hulls blast-cleaned, re-machined, and refinished, subassemblies cleaned and rebuilt, then the entire vehicle is reassembled on a production line.

In the vehicle rebuild activity, there are about 1.5 million square feet of covered area. This area includes two machine shops housing about $10 million worth of machine tools, including numerically controlled (NC/CNC) equipment, lifting devices up to 60 tons, automated bore, mill and drill machines that quickly restore correct precision holes in patched hulls, and automated shot-blast equipment that removes paint and scale from vehicle hulls in a single operation. Total value of production and other equipment exceed $147 million. Highly specialized support shops perform such tasks as electroplating, nomenclature plate making and silk-screen printing, fabric-leather-rubber fabrication and repair and electronic test and rebuild of weapons systems, including helicopter armament.

An automated storage and retrieval system directly supports the maintenance operations with parts and materials.

The Army's only roadwheel and track shoe rebuild facility is located at Red River. In the past 10 years, it has overhauled more than 2,000,000 track shoes and 500,000 roadwheels in our modern facility, for a savings of about $96 million. The addition of a fluidized bed process for removing rubber has enhanced its ability to increase production, while at the same time allowing it to reach objectives in reducing hazardous waste, volatile organic compounds emitted to the atmosphere, and utility costs.

Red River also has an elaborate electronics repair capability to support sophisticated systems such as the Multiple Launch Rocket System and Bradley Fighting Vehicle System and a variety of other missile support and aircraft armament subsystems. Many pieces of automated and semi-automated test equipment support these various programs.

DEPOT TENANTS

Red River Army Depot is host to 10 tenant organizations. The largest is the Defense Logistics Agency (DLA) Distribution Center, which has about 800 employees and houses approximately $6 billion in materiel for world-wide distribution. Red River has a special relationship with DLA, to which it is host, supplier, and customer. Unlike DLA activities that primarily serve the host installation, Red River's DLA depot does 80% of its business with other installations worldwide. DLA also operates a significant Defense Reutilization and Marketing operation that annually handles millions of dollars worth of materiel identified as excess or not economically repairable.

Other tenants include the Defense Finance and Accounting Service (DFAS), which is responsible for paying 28,000 civil service employees assigned to various Department of Defense activities. DFAS also provides accounting services to 116 Army installations. The U.S. Army Materiel Command operates a School of Engineering and Logistics, which provides training to supply, maintenance, safety, and engineering interns for all the U.S. Army. Also located on the premises is a Health Services Command clinic that provides medical and industrial hygiene support for Red River and its tenants. The clinic also provides medical support for a large population of active duty and retired military members and their dependents, as well as National Guard and U.S. Army Reservists who train at Red River. Another major tenant is the U.S. Army Test Measurement and Diagnostic Equipment Support Operation, which provides calibration support to the entire depot, the National Guard, Army Reserves, and the Federal Aviation Administration. The laboratory also operates an ammunition and small arms gage certification program for all areas west of the Mississippi River.

OUR REASON FOR BEING—OUR CUSTOMERS

Red River Army Depot's customers are many and varied. External customers include soldiers, project and commodity managers for the weapons systems, plus all of the subordinate commands within the Army Materiel Command that oversee Red River's specific industrial activities and processes. Support to these technically oriented customers includes assistance in design and preparation of technical data packages, parts procurement, fabrication, and program management. Red River also works closely with the soldiers in the field, listening to their needs and providing assistance teams to field locations when a personal level of service is needed. Due to Red River's multi-mission nature, depot personnel realize they are internal customers themselves. All support organizations provide products and services to the mission organizations, and other support organizations as well. Feedback from customers proves that Red River can react quickly and provide quality products and services through innovative methods at a distinct economic advantage.

OUR QUALITY JOURNEY BACKGROUND

Total Army Quality, the Army's adaptation of Total Quality Management, is Red River's management philosophy. Red River has used different concepts of quality management during the past decade. However, in June 1991, the Department of Defense implemented a competition program among all the services. The premise of the program was to achieve cost savings through efficiencies generated by workload competition. This change influenced and redirected Red River's view of the depot's future and intensified its transformation to Total Quality Management.

The Saturn Automobile Corporation, the benchmark for Red River's cultural change, was chosen because it has the type people programs, the type customer service, and the type results Red River wanted. Red River tackled change using the "7S" model (Shared Vision and Values, Strategy, Structure, Staffing, Skills, Style, and Systems).

Red River began its journey with an end in mind, a *shared vision*. In October 1992, the Executive Steering Committee comprised of depot directors, office chiefs, union chiefs, and special staff, met to develop Red River's vision. The words of Red River's 1991 nomination package for the Army's Communities of Excellence Award read "Excellence Begins With A Vision." It was agreed by all that Red River is an industrial complex and it must be competitive by excelling in the quality of products and services provided to its customers. From that came Red River's vision statement, "A Competitive Industrial Complex Excelling in Quality Products and Services." These words are codified in a picture, one that is easily understood by all members. Red River Army Depot is a team; therefore, all employees are referred to as members of that team. In order to be successful, all members of the depot team travel the road on a quality journey together with a single focus on Red River's vision.

The road to the vision is like a lens to focus Red River's efforts toward Total Army Quality. Red River collectively and aggressively built a participative, committed, and ready team capable of meeting the uncertainties of a constantly changing environment.

Leadership and communication are the keys to unlock the barriers Red River continually faces. Goals and objectives provide the path to follow with signposts along the way as additional support to help keep everyone headed in the right direction.

One of those signposts is "Quality of Life." Red River continually strives to provide a quality of life that is equal to or exceeds that of private industry. Red River understands that quality of life includes not just the facilities in which its members work and the tools and equipment provided to do their jobs, but most importantly, how they are treated. Red River doesn't just tell them they are important, it lets them know they are by providing many avenues of empowerment and involvement. Members are given the freedom to utilize their creativity and innovative ideas to the fullest extent and help them grow to realize their full potential. They are the experts and their contributions have a significant impact on the attainment of Red River's vision.

Changing the way Red River does business by redesigning its culture required commitment and personal involvement by senior management. The leaders are committed to the recommendations and principles of the National Performance Review (NPR). In fact, Red River implemented many of the review recommendations prior to the publication of the NPR report. Because of its leaders' strong commitment to continuous quality improvement and a new way of doing business, Red River is recognized as a model for implementing change, training, empowerment, and union management partnerships. For example, Colonel Thomas A. Dunn (Commander, July 1992 to July 1993) and Jim Taylor, head of the largest union at the time, addressed the National Partnership Council on May 11, 1994. Colonel Dunn was also recognized in a leadership article published in the July 1994 issue of *Government Executive* magazine. Colonel Dunn was the Commander when the depot began actively pursuing its quality journey. As each new Commander has taken the reins, he has continued to reinforce the importance of staying on that journey. Not only have Red River's Commander and top union leader been champions of the quality journey, but other top leaders, including the Civilian Executive Assistant, Directors and other supervisors, and union officials have spoken to a large number of federal and private sector audiences to share Red River's process for forming union and management partnerships and the successes of its overall quality journey.

Just as important as a shared vision are shared values. "HEARTS" is an acronym that represents a trait that identifies Red River's individual and organizational values (Honesty, Ethics, Accountability, Respect, Trust, Support).

These behaviors, when consistently followed in an organization, empower people to be all they can be. They help team members to consistently exceed their own and others' expectations. *Our value system promotes a continuous improvement attitude constantly focused on our customers.* Aligning vision and values required a cultural change.

Red River's "HEARTS" values are tied to its "HEARTS" training program, a learning process that challenges individuals to go beyond their perceived boundaries, encouraging them to work with others to solve problems and attain their goals. Individuals learn to view obstacles "as opportunities for growth in an atmosphere that is fun, supportive and challenging." Red River's ropes course is in an outdoor setting covering approximately 10 acres nestled in the woods near the depot's recreation area, known as Elliot Lake. It focuses on developing human relationships within a group, gives new meaning to the importance of working together as a team, and challenges each person as an individual. The goal is to provide a catalyst for personal, professional, and educational growth.

Why "HEARTS" training? Just saying "We need to change" does not make change happen. Red River needed the best training available that would help everyone understand and better accept a new environment that was destined for constant change. Red River had to make a dedicated and focused effort to pull together as a team and move from placing blame on its neighbors and management or the union every time something didn't go the way it wanted it to go. It also had to embrace a concept that builds excellence through total member involvement, empowerment, and continuous process improvement. This new concept departs from the old way of listening to only a few thinkers at the top, and recognizes that there are hundreds of doers who are, in fact, the experts of their processes and deserve to be heard. In order to do this Red River had to work very smart and very quickly to do away with old paradigms, such as, "The people turning the wrench only know how to turn a wrench," and "If a specific division is not meeting production, we have to get the people to work harder." The people are not the problem and just working harder is not enough. The processes are what need fixing, and Red River needs to work smarter, not harder. More is not better; better is better.

Red River developed a strategy to make Total Quality Management its new management philosophy and provide its customers with the highest quality products, on or ahead of schedule, and at the least possible cost.

To accomplish these strategic goals, in 1991 the Executive Steering Committee, with the help of several Process Action Teams developed a new *structure.* Two layers of supervision were removed, percentage of direct labor positions increased, base operation positions decreased, inspection functions moved into the mission areas, and supervisory positions reduced by 49%. Since 1991 the member to supervisor ratio has changed from 9 to 1, to a current ratio of 16 to 1.

The new organization required the reassignment of hundreds of personnel. In order to ensure fairness, it was staffed using a fully automated system with the unions and members actively involved. They were given detailed information throughout the process and verified data accuracy. Due to government downsizing, Red River has undergone several reorganizations since 1991, and through each one, its union partners have been thoroughly involved in the implementation process—a great departure from the past when they were only involved after they filed appeals.

Red River developed new *skills*. All members have attended "HEARTS" training, the foundation of the change process, as well as "Customer Service Training." Hundreds have attended Steven Covey's "Seven Habits Training," and several hundred have received 40 hours in Windows computer training. *In 1991,the average training hours per depot member was 17; in 1994 that number was increased four-fold to 71.* Training continues to be a viable tool in remaining current and competitive in today's changing Army.

Along with providing new skills, Red River changed its style. Its union and management moved from the old adversarial way of doing business to the new way of cooperation, collaboration, and co-ownership. These partnerships provide the support necessary to make change happen.

A unique partnership exists at Red River between management and all six unions. The largest union until recently, National Association of Government Employees, was successfully challenged by the National Federation of Federal Employees. The Commander has already established a relationship with the new officials and has paved the way to a continued partnership. The new union, which saw how successful Red River's previous partnership was, is continuing the trend of a working partnership.

Senior management's invitation for union presidents to participate in all aspects of depot operations was the catalyst for gaining union support and cooperation. Union representatives actively participate through membership on Red River's Executive Steering Committee and attendance at all staff meetings. Partnerships extend beyond senior management and union chiefs. They exist at all organizational levels with supervisors and union partners managing the business together. Red River has moved from the old to a new way of doing business. The results have been outstanding. Supervisors and union representatives work to resolve member grievances rather than invoking arbitration. When agreements are reached, the partnership is strengthened and tax dollars required for arbitration are saved.

Managers and supervisors are leaders, mentors, coaches, rewarders, and resource providers to Red River's Self-Managed Work Teams and Process Action Teams.

Finally, Red River is installing *Systems* that support its Shared Vision. Group performance appraisals and recognitions are increasing every day. Red River did away with individual annual cash performance awards. At the end of the fiscal year, each member receives the same or no performance cash award based on the Net Operating Result (revenue less expenses). Members celebrate successes together. A picnic was the way members wanted to celebrate winning the Quality Improvement Prototype Award in 1995. In addition, every member was given a shirt reading "Together We Stand Proud" on the front and an eagle on the back carrying a blue ribbon embossed with "Quality Improvement Prototype Award Winner."

IMPROVEMENT METHODS

Assessment

You must know where you are before developing a plan for your journey. Before accelerating quality efforts in 1992, Red River conducted an in-depth self-assessment. Since then, it has continued this process. In 1992 and 1993 Red River used a guide developed by the Army Materiel Command patterned after the President's Quality Award. The Federal Quality Institute's Self-Assessment Guide was used in 1994. These assessments provided valuable feedback and were used to identify specific areas requiring focus. Continuous improvement resulted in steady progress each year. The Malcolm Baldrige National Quality Award criteria was the basis for self-assessment in 1996, 1997, and 1998.

Information and Analysis

Information empowers. *Communication* is a word used and practiced daily among all members. Decisions are based first on fact, then whether or not they are in line with Red River's vision, quality values, internal/external customer requirements, and last, but not least, what is best for members.

One way Red River communicates the importance of increased customer focus and quality values is by rewarding those who make contributions. Members are recognized for their involvement in a variety of ways. Rewards include "HEARTS" and "Total Army Quality" Commander's coins, Quality Champion T-shirts, "HEARTS" certificates, and even time off from work.

Additionally, customer focus and quality values are frequently communicated utilizing a variety of methods, such as the monthly depot newspaper, letters from the Commander, monthly member coffees with the Commander, and weekly "FYI" programs taped by the Commander and broadcast over the Local Area Network, to name just a few.

All of Red River's measures for analysis are developed based on a single focus—the vision of a competitive industrial complex excelling in quality products and services. To fulfill its vision, Red River's emphasis is on its internal and external customers. It develops measures that

- First, determine whether or not customer satisfaction will be enhanced by improving quality, increasing efficiency and effectiveness, decreasing cost and cycle time, and reducing waste
- Second, identify the relationship of that measure to Red River's goals, objectives, and annual business plans to ensure that it measures the important things
- Third, make certain the activity or process is measurable

Key types of data collected relate to customer satisfaction, member involvement, training cost, schedule, and product/service quality and performance. These six major areas of measurement support the execution of Red River's strategic quality and operational business plan. Data collected for monthly review and analysis of quality, operational performance, and business plans comes from every organizational area on depot.

- **Customer satisfaction/complaint data** to continuously identify improvements and current/future requirements
- **Involvement**, since members are our greatest asset
- **Training** to ensure all members are provided empowerment and teamwork skills
- **Financial data** to control revenue and expenses
- **Production schedules** to analyze effectiveness
- **Performance information** to review efficiency, productivity, and waste reduction

Daily production run charts are kept at cost center level. Information is then summarized in standardized charts to ensure effective communication and analysis by all levels. Members of Red River's Process Action Teams and Self-Managed Work Teams collect and analyze data for specific process improvements.

Management of Process Quality

In October 1993, inspectors previously located in the Quality Assurance Directorate were assigned to and are now working right along side the members in the maintenance production mission areas. This combines a proactive and preventive approach for quality assessments with the continuous process improvement assessments of performance, an important aspect of Red River's new way of doing business. By encouraging its skilled members to work together as partners and teams, the quality and performance of Red River's processes are improving. The planning and implementation of this major transition was successfully accomplished by Process Action Teams that continue to analyze and improve this new structure, a concept that provides direct combination and correlation of quality and performance.

Quality and Operational Results

To gain quality and operational results, Red River utilizes its greatest resources, its members. You can have the highest technology, the best machinery, and state-of-the-art processes and procedures, but if you don't pay attention to your members, if you don't really care about the people that do the job, you will not be successful. Red River's results stem from its cultural change.

Red River is doing more than just providing its customers high quality products and services, more than increasing productivity with decreased operational costs; it is creating a high performance team in which the members not only care about customer satisfaction, productivity, and costs, but also care about each other and about changing the negative perceptions of our government. Red River's results indicate how successful it is at doing this.

In addition to focusing energies on process improvements in mission areas, Red River also places emphasis on improving the procedures of its internal support functions that significantly impact the success of its missions.

The members of Red River Army Depot are number one and are treated with the respect and dignity they deserve. When the employees of an organization are *valued* as important members of the overall team, they in turn, make exceptional *team* players. They strive to always do their part to get the results required to meet or exceed the expectations of their customers.

Customer Focus and Satisfaction

In 1978, Red River Army Depot adopted the motto "OUR BEST—NOTHING LESS." This was a conscious effort to highlight and emphasize a philosophy and service standard that has been in effect at Red River Army Depot since its beginning in 1941. The motto is used in correspondence, on signs, letterheads, decals, and in special promotions of events.

The motto carries with it two messages. One message is to the members of the Red River Army Depot workforce—we will do our best in everything we do. The second message is to Red River's many customers—it tells those customers to expect and anticipate Red River to provide only the best, nothing less—the best in workmanship, appearance, functionality, quality, and cost. The four little words make a giant statement regarding Red River's commitment.

Today, Red River finds the motto still remains with it. However, it also finds it has been joined by other statements and practiced philosophies that communicate Red River's local commitment to its customers. For example, Red River's customers enjoy special treatment with designated Customer Parking. Spaces are marked "Customer" and located directly in front of Red River's buildings. Parking policy was changed not for our customers, but also for members. In addition to special customer parking, the only other marked spaces are for the handicapped and official government vehicles. The statement here is that *no one person on the depot is more important than another.*

The success Red River Army Depot enjoys in establishing and maintaining good relations with its customers is primarily centered around *communications and producing quality products and services.* An analysis is made of the specific activities that bring Red River into contact with the customer most often, and then condensed into one of these two important general categories. The application of these two principles is the foundation for the many partnerships that have been formed over the years with both internal and external customers. The objective is to move from maintaining the basic relationship to a higher level of *partnership.* Assessing Red River's products and services for quality requires the examination of many, many specific and detailed processes. Customer satisfaction is largely determined by how well Red River performs.

Red River Army Depot members are proud of over 50 years of service and support provided to soldiers, the Department of Defense, and the nation. It feels that it is the best at what it does. It knows that it has not reached the pinnacle of its vision. It is on a continuing journey to its goal; however, its present performance relative to that of similar providers in the public and private sectors is exceptional. Over the years, its customers have demonstrated by their words and their actual confidence placed in Red River that this is true.

STATUS OF QUALITY REPORT

Red River continues vigorous implementation of Total Army Quality drawing on the creative genius of all of its members. In order to sustain a competitive advantage, people and quality come first. Visual evidence of its commitment to Total Army Quality principles includes Self-Managed Work Teams throughout the depot to empower members and encourage ownership, and placing union partners alongside managers at every level of supervision to work together with a shared vision and shared values. These steps are changing its culture and making it responsive to changing requirements while making it highly competitive in an Army environment that is reducing both budgetary dollars and manpower.

Red River understands that *"even if you are on the right track, if you sit still, you will get run over."* Its journey continues as it remains focused on its goals and objectives.

RESULTS OF INITIATIVES

Process Action Teams; Self-Managed Work Teams

Red River's Process Action Teams work to improve a specific process and, once they complete the task, disband. Over a period of 7 or 8 years, more than 130 Process Action Teams have been formed, with 70% of those being cross-functional.

In October 1993, it reorganized for quality, creating an environment conducive for Self-Managed Work Teams. Layers of supervision were reduced from five to three. Teams are formed within their own functional areas. Typical team responsibilities include determining job assignments, planning work, approving leave, identifying needed training, resolving team conflict, initiating disciplinary action, developing job descriptions, developing performance standards and conducting team appraisals, and recommending persons to fill vacancies. Red River now has 60 Self-Managed Work Teams with membership representing approximately 30% of the population.

Model for Change

Red River has shared its approach to Total Army Quality with more than 6,000 people outside its own organization. Red River is a nationally recognized leader for cultural change.

National Performance Review (NPR)

Red River members are dedicated to the principles of the administration's National Performance Review. The September 1994 NPR Status Report included concepts and successes of its quality strategy.

Partnerships

New partnerships have been formed with customers and suppliers, and among members. Partnerships between union and management exist at virtually all levels of the organization. As a result of the new way management and union does business, *all types of grievances have decreased by approximately 80% at an average annual savings of $600,000.*

President's Quality Award

Red River Army Depot applied the first time in 1993 for the Federal Quality Institute's prestigious Quality Improvement Prototype Award. It was selected as one of only nine finalists within the federal government. Although it did not win the final competition, the participation by the depot and its members brought many benefits relating to quality improvements and emphasis on product and life quality. The criteria of that award provides a blueprint for Red River's quality journey. The evaluation process provides a thorough outside assessment of Red River's progress. The benefits of the Federal Quality Institute's Quality Award Program lie in the pursuit. By using the feedback from the assessment, Red River was able to institute changes and in its second attempt at the competition, it were successful in becoming a winner.

Mission and Mission-Related Results

Red River's members are empowered to get results. They all work toward a single vision, *A Competitive Industrial Complex Excelling In Quality Products and Services,* and equal annual cash awards. The only annual cash performance awards are based on annual Net Operating Result (revenue less expenses). Due to the productivity improvements and waste reductions accomplished by depot team members, Red River has been able to award its total workforce with a NOR payout for several years. It ended Fiscal Year 1997 $11.3 million better than plan! Each member received a $500 cash award.

Examples of specific mission and mission support results follow.

The Multiple Launch Rocket System Project Office originally intended to use the manufacturer of the Electronic Unit to apply a modification to 360 units. However, Red River was given the opportunity to make a proposal. It was assigned the modification in lieu of the contractor. As a result of better quality, cost reduction, and shorter program execution time, the complete project was awarded to Red River. Units were completed quicker than projected by the contractor at a *savings of $59,000,000.*

Red River competed with a contractor to perform the design work and conversion of the tow M113A2 Army Personnel Carriers to the M configuration. Red River was awarded the project and completed the two vehicles 2 months ahead of the 6-month schedule and approximately $200,000 *under budget.*

Red River produced three XM1101 Smoke Vehicle prototypes 3 months sooner than a contractor had committed, and at a cost that was *approximately $1,500,000 less* than the contractor's proposed cost. Red River completed the design work, conversion, and acceptance of the three vehicles *ahead* of schedule and below budget.

Improvements in Red River's mission processes are continuous. For example, its Electronic Division was behind in the Production Schedule for Bradley Fighting Vehicle turrets during the first and second quarters of fiscal year 1993. During March 1993, vehicles remained to be mated with turrets originally scheduled for November 1992 through February 1993. Simultaneously it was trying to "ramp-up" to 18 conversions per month. Knowing "if you always do what you've always done, you'll always get what you've always got," the division approached this problem with a new outlook. The members came up with a new process and implemented the improvements. This resulted in a more supportive environment with a "team concept," as opposed to individual efforts in the "traditional concept." The results were astounding! The backlog of turrets to be worked changed to a backlog of turrets ready for use. Personnel requirements have decreased by approximately 25%, and morale has improved drastically.

Quality Deficiency Reports are customer-generated reports that indicate problems with products received from Red River Army Dept. *None have been received in the Ammunition Operations Directorate since fiscal year 1990.*

Red River has many successful teams; however, its "Super Crew," a self-managed work team in its Ammunition Directorate, has been a leader and a "model" for the formation of other teams. They actually set up and created their own operation from scratch based on the requirements of their customers. They studied the Technical Order to learn the procedure for test Maverick missiles. As a team, and without supervision, the Crew repaired and modified their test equipment, and created a Standing Operating Procedure. Once they initiated this new function, they continued to improve the process *and increased their production from 3 to 15 missiles per day.* This team ultimately received Vice President Gore's prestigious Hammer Award.

As a result of the work done by the Electronic Data Interchange Process Action Team, the 79 offices throughout the depot that routinely generate Requests for Issue of supplies and equipment have a computerized alternate to filling out forms by hand. The issue form is used to order specific items and start the supply process. The team realized that if they could improve the accuracy of the information on the form and the way the form is handled, then the entire process could be improved—*yearly savings are estimated at more than $25,000.*

Red River's "Credit Card" process action team was formed to look into why procurement credit cards were not being extensively utilized to improve receipt time for internal customers. It found that cardholders purchased an average of only one item per month because the processing requirements were logistically complex and labor intensive, and cardholders could not see the benefits. The procedure was slow and frustrating. The team objectives included improved customer satisfaction through increased response time, decreased processing costs, and improved partnerships with suppliers. The team re-engineered the process resulting in *decreased average customer receipt time from 120 days to only 5 days with a conservative estimated first-year procurement processing savings of more than $1,000,000.*

In addition to continuously improving upon its initial quality concepts, Red River is committed to a new major challenge for the depot: Contractor Performance Certification Program (CP)[2]. This new initiative is an Army Materiel Command certification program that promotes Quality Achievement within the U.S. Army Contractor and Depot Support Communities.

The (CP)[2] program seeks to establish the Quality System Management outlined in the International Standardization Organization (ISO) 9000 series of requirements. This effort entails a teaming between the depot, its customers, and the Industrial Operations Command (our next higher headquarters). In a nonadversarial environment, these entities team to improve the depot's processes until the criteria are met with confidence of continual improvement. Benefits include improvement of product quality, reduction of cost, increased productivity, enhanced readiness, and greater customer satisfaction. Red River received certification in its ammunition mission area in July 1998 and expects certification in the other two areas to follow soon.

Another direction for the future of Red River Army Depot is to increase its network and partnerships with world-class private companies. It is focusing its energies on benchmarking with a comprehensive plan currently being developed by a Process Action Team. All of its members will be involved in order to capitalize on the effort and ensure maximum benefits.

Red River will continue to travel its journey with the dedication and perseverance of a high-performance team. Its managers will continue to become more and more skilled as coaches and mentors nurturing our self-managed work teams as each grows to full potential.

- *Begin with an end in mind*—Develop a vision and align goals and objectives for attainment. Continually communicate that vision in a variety of ways to members, customers, and suppliers. Remember, energy follows focus.

- *Identify individual and organizational values*—Once those values are identified, commit to changing the culture of the organization. This commitment must begin with top management, but the same commitment is required of middle managers as well as all the members of the organization. Organizational change begins with individual change. Invest in training to prepare everyone for success.

- *Strengthen ongoing partnerships and develop new ones*—Lay a strong foundation of partnerships between union and management. Then build on that foundation a support system that cannot be broken. Get rid of the old baggage that may have been around for years by joining together and making decisions based on what is in the best interest of the members and the organization as a whole.

- *Include customers and suppliers as partners in your quality transformation*—Establish customer service standards based on the expectations of your customers, rather than what you think those standards should be. Emphasize the quality requirements for the products and services of your suppliers. Always strive for win–win situations.

- *Celebrate*—Regardless of whether you eliminate a major barrier or just manage to get around it, celebrate. Even the highest mountain is climbed one step at a time. *Celebrate each step.*

LOOKING BACK

The quality journey is a difficult one. It is important to routinely stop, breathe, and refocus your efforts. As Red River looks back at how far it has come, it is clear that Red River's quality journey is like a puzzle coming together. The implementation of just one concept is not enough. Red River focused on its vision and its values, and aligned all of the pieces of the puzzle below to produce the maximum positive impact on its depot's culture and performance—organizational strengths that when maximized ensure competitive vitality.

The first important piece of the puzzle is commitment. The piece which marks significant progress is results. However, all of the other pieces must be used and integrated as you travel your journey. The criteria for the Malcolm Baldrige are like a road map which, when followed closely, will help keep you from making wrong turns or taking short cuts which could add unnecessary miles. Focus on quality of life of your members and the satisfaction of your customers. Prepare your passengers for a successful journey through training, and reward their contributions so they will be encouraged to continue being creative and innovative. Design an organizational structure conducive to teamwork, Process Action Teams (PATs) and Self-Managed Work Teams (SMWTs), and establish focus groups at the top to nurture and provide the required resources for teams to succeed. Build partnerships throughout the organization with your customers. Benchmark best practices with cutting edge organizations. Share your strategies and successes with others and strive for those initiatives identified by the National Performance Review Team.

This puzzle is not an easy one to put together. However, if you want to be a *world-class organization*, it is a puzzle you must attempt.

COMPANY

U.S. Army Armament Research, Development & Engineering Center (ARDEC)

PRODUCT/SERVICE:

Delivery of quality weapon, munitions, and fire control systems and associated materiel. Munitions, guns, fire control and associated weaponry.

LOCATION:

Picatinny Arsenal, New Jersey

NUMBER OF EMPLOYEES:

3,058 employees

KEY PERSONNEL:

Brigadier General John P. Geis
Technical Director: Michael Fisette

Case Study Contributors: George Cherenack, Kathryn Daut (TQM POC), Ellen Haveman, Donelle Denery, Glenn Caltabilotta, Tom Zalasky, Jamie Ruffing, Vince Stenziano, Geza Pap, Dave Murawski, Mark Eldridge, Beth Albinson, Joann Scachetti, and Jim Matanin

KEY USES OF INFORMATION:

■ Indicators
■ Benchmarking and competitive comparisons
■ World Wide Web

ORGANIZING FRAMEWORK TO CREATE KNOWLEDGE:

Baldrige-based criteria including the President's Quality Award (PQA) Program criteria, Army Performance Improvement criteria, and NJ Governor's Award for Performance Excellence criteria

AWARDS AND SIGNIFICANT RESULTS:

■ Army R&D Excellence 6 times (1989, 1990, 1991, 1996, 1997, 1998)
■ Army R&D Organization of the Year 1986 and 1995
■ 1995 President's Quality Improvement Prototype Award
■ 1996 Presidential Award for Quality (government equivalent of the MBNQA)
■ 1996 Army Communities of Excellence Award (Army's MBNQA equivalent)
■ 1996 New Jersey Quality Partner Award
■ 1998 New Jersey Quality Achievement Award (re-named Governor's Award for Performance Excellence)
■ 1999 Chief of Staff, Army (CSA) Army Communities of Excellence (ACOE) Award

INTRODUCTION AND BUSINESS OVERVIEW

The United States Army Armament Research Development and Engineering Center (ARDEC) has traveled on its quality journey since 1988. It has listened to the *Voice of the Customer* to create an organization that is strongly customer focused and which delivers products and processes of high quality as measured by the customer and the criteria of the President's Quality Award. It has achieved this level of performance with reduced cost and cycle time based on a culture of team work and application of the precepts of total quality. It has repetitively applied the Baldrige-based criteria through the President's Quality Award criteria, the Army Performance Improvement criteria, and the New Jersey state criteria to systematically evaluate and improve its management practices and operational performance. It is proud to be recognized as the winner of the 1996 Presidential Award for Quality, Army R&D Excellence six times, R&D Organization of the Year 1986 and 1995. Also, the 1995 President's Quality Improvement Prototype Award, 1996 Army Communities of Excellence Award (Army's Baldrige equivalent), and the 1996 New Jersey Quality Partner Award.

ARDEC provides the U.S. military with the firepower necessary to achieve decisive battlefield victory. It satisfies customers through timely and cost-effective delivery of quality weapon, munitions, and fire control systems and associated materiel that meet or exceed customer needs. ARDEC engineers execute and manage life cycle engineering processes required for the research, development, production, field support and demilitarization of ammunition, weapons, fire control and associated items. As a world-class R&D Center, ARDEC was selected to be the Army's executive agent for research and development of pollution prevention.

For almost a decade, ARDEC has been challenged by significant reductions in funding and human resources. Despite this, continuous improvement enables ARDEC to increase the quality of its products and services and, simultaneously, to reduce cycle time and costs. Of special significance is the importance of communications and customer focus. Every activity within the Center involves communications. When problems arose in the past, root-cause analyses often showed ineffective communications as the cause. ARDEC learned that effective communication requires a strong team approach, continuous fact-based evaluation, and constant employee effort directed toward improvement.

Leadership is essential to innovation. ARDEC has learned that cultural change via training and leadership by example are critical to establishing a customer focus. Cultural change is difficult to introduce and to sustain. Consensus among leaders on values, mission, vision, goals and objectives is an indispensable first step and difficult to achieve. ARDEC is fortunate in the succession of top leaders who shared a common vision and who built upon past successes to achieve constancy of purpose toward the improvement of its products and services. This has created a leadership culture where all levels focus on customer satisfaction. The leadership practices of senior executives provide the impetus for the advancement of ARDEC's quality culture.

Strong common purpose and communications among senior leadership, major customers, and union leaders is fostered through participation on ARDEC's TQM Executive Council. To strengthen major customer participation, the Commanding General created the position of PM Advocate in 1991 which evolved into an overall Customer Advocate to serve as a single point of contact to facilitate customer satisfaction. The Customer Advocate participates in the TQM Executive Council meetings and ARDEC's regular metrics forum, the Systems Measurement Review. This customer knowledge led ARDEC to adapt this approach to other customer segments. All of the union leaders are members of the Executive Council, providing insights into the process and partnering on all significant changes.

Deployment of the ARDEC vision, values, and goals is a matter of constant leadership attention as evidenced by their dedicated time, attention to process, and deep caring for people. The leadership uses many methods to strengthen the work force knowledge and commitment to quality and customer satisfaction. The ARDEC business planning process is steeped in quality approaches and sown throughout the work force, customers, and suppliers. Key to high levels of quality performance and measurement is training. Training is accomplished from the top down and focuses on increasing awareness of total quality management principles and practice. At the top, the need and the benefits of cultural change from a traditional, hierarchical, bureaucratic, organizational environment to an adaptive, customer focused environment are emphasized. All members of the workforce are involved in problem-solving efforts and contribute to improving work processes. All managers receive training on quality management principles and practices. General quality and cultural change training is conducted repetitively to sustain workforce awareness of the importance of quality management principles. Significant numbers of ARDEC's people have been trained in the application of Baldrige and PAQ criteria.

Teams are the backbone of the Center's quality and productivity improvement effort: *ARDEC is teaming!* ARDEC makes a significant investment in the selection and training of team facilitators. This investment pays dividends in terms of team effectiveness and contributes greatly to teams' results. Establishing a customer focus required close examination of each work process and entails new behaviors. Although many of the Center's process improvement teams were initially chartered for a short-term process improvement project, many of the teams continue to monitor the results of their changes and meet to ensure the gains are sustained and opportunities for additional improvement are identified and exploited. For many process improvement projects, the team owns the process and is empowered to implement the needed changes. ARDEC invests heavily to train individuals as experts to meet specific new knowledge and skill requirements.

ARDEC has formal measurement guides and practices in place to monitor metrics based on Baldrige and PAQ criteria. Determining what to measure and how to measure it is a challenge that will never end in a dynamic ever-changing pursuit of quality. A Systems Measurement Review process drives improvements to the Center's systems. Benchmarking the best is a formal part of ARDEC's measurement process. Change is based upon clear active customer communications and an understanding of what is important to customers—all of which is linked to operational performance and process improvement measurement over time against established benchmarks.

Recognition of ARDEC's hard work to improve quality is gratifying. It reflects great credit upon all the people of ARDEC. Designation as a President's Award for Quality winner confirms ARDEC's management approach and provides the encouragement to continue the quality journey.

MISSION

ARDEC is the U.S. Army Armament Research, Development and Engineering Center. It provides America's soldiers, sailors, marines, and aviators with unit and individual weaponry needed to achieve decisive battlefield victory. ARDEC performs quality focused armament life cycle science, engineering, and technical support. It is also the Army executive agent for research and development of pollution prevention. Its mission is to execute and manage life cycle engineering processes required for the research, development, production, field support and demilitarization of ammunition, weapons, fire control and associated items.

ORGANIZATIONAL STRUCTURE

ARDEC is located just north of Interstate 80, approximately 40 miles west of New York City, at Picatinny Arsenal near the town of Dover in Morris County, New Jersey. It reports to the U.S. Army Tank-Automotive and Armaments Command (TACOM) in Warren, Michigan. TACOM, in turn, is a major subordinate organization of the Army Materiel Command located in Alexandria, Virginia.

ARDEC is autonomous with a full range of strategically aligned mission, base operations, and staff elements. The Center employs approximately 2,648 people at Picatinny Arsenal. Another 410 ARDEC employees work at the Benet Laboratories at Watervliet Arsenal near Albany, New York, and other co-located elements at Rock Island Arsenal, Illinois, and at various other satellite locations throughout the country. Most of the Center's work is of an extremely technical nature; more than half of the workforce are scientists and engineers, many with advanced degrees. The Center also has approximately 40 military personnel assigned. The Department of the

Army civilians are represented by four unions: the National Federation of Federal Employees, the American Federation of Government Employees, the Fraternal Order of Police, and Federal Uniformed Firefighters.

PRODUCTS AND SERVICES

ARDEC's key products are munitions, guns, fire control, and associated weaponry:

- 55 advanced technology products
- 85 products are in design and development
- 151 products are in production
- 1,386 products supported in the field

ARDEC provides the nation's expertise in engineering and scientific services for armaments and munitions development. Its 1,600 engineers and scientists are the world's premier experts in

- ammunition
- weapons
- fire control systems
- fuzes, lethal mechanisms, and warheads
- demolition munitions
- mines, bombs, grenades, and pyrotechnics
- explosives and propellants
- practice and training munitions
- pollution prevention

The high-performance workforce provides customers and suppliers with simulation, virtual prototyping, advanced scientific computing, and a wide, sophisticated range of engineering services to include robust design techniques, statistical process control (SPC), and Quality Functional Deployment (QFD).

OUR PERSONNEL ARE LINKED TO KEY CUSTOMERS

- Picatinny Arsenal, NJ, ARDEC's largest location, is shared with the Program Executive Office (PEO) for Ground Combat Support Systems (GCSS), its Project Managers (PMs), and other PMs. They are responsible for the material acquisition (cost, schedule, and performance) of major weapons systems and items of equipment for the armed forces. This group of customers represents a major source of revenue for the Center. Since the acquisition process includes concept exploration, engineering, and manufacturing design and production, it is critically important that ARDEC engineers, scientists and technicians completely understand the PEO's, PM's, and the soldier's needs.

■ Watervliet Arsenal, NY, a component of the Army Industrial Operations Command (IOC), is the location of ARDEC's Benet Laboratory, which provides direct engineering and scientific support to the Arsenal, in the design and manufacture of gun barrels and associated equipment.

■ Rock Island Arsenal, IL, is where ARDEC provides direct support to the Industrial Operations Command (IOC) and the Army Armament and Chemical Acquisition and Logistics Agency (ACALA), including the Arsenal's production missions. As the Army's weapon manufacturing entity, the IOC produces the huge amounts of conventional ammunition required by the Department of Defense. ACALA does equivalent work for weapons and associated products. ARDEC and these organizations all work with defense contractors to meet the U.S. armed forces' needs. ARDEC provides this customer with product technical data in the form of detailed drawings and specifications and then supports production with engineering services.

■ ARDEC employees are physically co-located with customers, such as the Battle Labs, and with suppliers, such as the Army Research Laboratory. Army schools and Battle Labs that are part of the U.S. Army Training and Doctrine Command (TRADOC) are focused on designing and developing future Army organizations to accomplish combat and non-combat missions. Army schools are representatives of ARDEC's ultimate customer; the Soldier. These organizations are inherently horizontally focused across battlefield operating systems and are total team efforts. Grounded in the lessons learned from recent world-wide operations and the expanding opportunities provided by computer-enhanced technology, the Battle Labs and ARDEC conduct joint Advanced War Fighting Experiments to establish and empirical basis for requirements and investment recommendations. These experiments emanate from futuristic concepts and employ an Integrated Product Team approach.

SUPPLIERS

ARDEC's suppliers include approximately 3,000 development and production contractors with which it works to develop and produce armament and munitions. In addition, there are approximately 7,000 contracts awarded yearly for base operation services. Another major supplier, the

Army Research Laboratory (ARL) at Adelphi, Maryland, other government laboratories, and acadamia provide basic research for ARDEC development programs. ARDEC is constantly improving relationships with all of its suppliers through contractor performance certification programs, Commander/Contractor reviews, government/industry workshare teams, and partnering initiatives.

MAJOR ISSUES

The armament industry involves both the public and private sectors. Due to the nature of ARDEC's work, safety is the critical constraint on its lifecycle engineering processes.

Over the last decade, the U.S. military profoundly changed the nature of its mission and combat roles. The global political and economic systems that were established and maintained for decades have changed. As a result, DoD is experiencing a period of deep budgetary cuts.

ARDEC must ensure that it continues to be the best value supplier of armament research, development, and engineering services to its customers. To improve quality, ARDEC is enhancing its partnerships with its customers, suppliers, and its employees while working to streamline processes and structure.

OUR QUALITY JOURNEY

Dr. Walter A. Shewhart, the "Father of Quality Control," was a consultant to Picatinny Arsenal in the mid-1930s at the request of First Lieutenant Leslie E. Simon. Simon was one of the Army's earliest and most influential advocates of statistical methods of quality control. He directed the preparation of a manual entitled "Instruction for Control of Quality through Percentage Inspection," which incorporated many of Dr. Shewhart's ideas. These ideas made a significant contribution to the war effort during World War II. Simon expanded this shop manual into "An Engineer's Manual of Statistical Methods." This book was instrumental in spreading Dr. Shewhart's ideas to both war plants and engineer and production manager classrooms. It became the basic text for quality control inspection of ordinance items.

During the 1940s and 1950s, Picatinny focused on manufacturing. It used statistical quality control as an in-plant tool for product sampling, but the emphasis was on getting the product delivered. During the 1960s, quality control became quality assurance and emphasis sifted to conformance to specifications. Many product enhancing techniques came into use during this period. They included zero defects, value engineering, producibility studies, SPC and fault tree analyses. In the late 1960s and early 1970s, quality focus expanded to include quality of design. Recognizing the importance of design factors, Picatinny developed tools to track reliability and performance. Risk management, Program Evaluation, and Review Technique (PERT) and root cause analysis were in use at Picatinny.

When manufacturing operations ceased at the Arsenal in 1977, the principal work performed at Picatinny shifted to research, development, and engineering. Production shifted to other government facilities and industry. This dramatic change in the nature of the primary work of the Arsenal gradually expanded management's focus from quality assurance to product and process improvement during the 1980s (product assurance).

In today's environment, ARDEC must provide long-term reliable products, low logistic costs, and new high technology features, while offering customers a range of choices, and delivering continuously improving product performance and excellent service. ARDEC has committed itself to accomplishing this transition through the precepts of total quality.

During the late 1980s ARDEC's executive leadership embraced total quality management and took responsibility for quality. The Deming philosophy took hold. ARDEC began to broadly apply his appreciation for systems and process. It combined its understanding of statistical theory and the impact of variation (learned in high volume ammunition production) with a rediscovered respect for individual talent, experience and creativity, and a renewed appreciation for customer requirements and expectations. ARDEC's drive for total quality shifted to a focus on cultural change that addressed process improvements in all technical areas and included business and support service processes, employee involvement, and customer satisfaction. ARDEC began by initiating employee involvement efforts using quality circles, middle management working groups, and *"power down"* initiatives.

In 1988, a cross-functional team visited Air Force Systems Command, Defense Logistics Agency, AT&T, and MIT to investigate their deployment of Total Quality Management (TQM). Based on these visits, ARDEC launched its formal TQM effort.

Brigadier General (BG) Joseph Raffiani strove to create a quality and customer focused environment. In 1989, he invited the Defense Systems Management College (DSMC) to train ARDEC top-level management in TQM in off-site sessions. BG Raffiani challenged ARDEC to begin TQM as a process. He provided the leadership to begin the transformation into a world class organization. The training was followed by a series of sessions which focused leadership on the quality journey. At this point,

ARDEC began a tradition known as *"passing the quality sword"* when one commander replaces another.

BG William Holmes accepted the quality sword from BG Raffiani and dedicated his leadership energy into TQM thrusts. The DSMC instructors responded to BG Holmes' request, and in 1990 held an additional off-site workshop for senior executives. This workshop resulted in the formation of the TQM Executive Council (EC). The EC met every 6 weeks to develop ARDEC's Vision and established its original Strategic Goals to steer quality improvement. In 1991, the EC recognized the need for an architecture to systematically plan, organize, and measure improvement activities. The Commanding General and Technical Director led the EC to the identification of our Systems and System Owners that set the strategic direction and built the leadership system for its business processes. Systematic process improvement activity takes place within this framework. Organizational directors are members of the EC. They use information and analysis to build quality throughout ARDEC and ensure active participation in cross-functional improvement efforts. In 1991, the EC changed its meeting schedule to every 2 weeks to keep pace with ARDEC's quality road map. During this period ARDEC adopted the AMEC integrated model for TQM education.

In 1992, BG Harvey Brown assumed Command and adopted Management Through Leadership (MANTLE), to create a listening and learning culture. He instituted Concurrent Engineering (CE) and expanded upon the CE concept via Integrated Product Teams to build upon our leadership system and strengthen our capabilities. To provide a benchmark for managers and more specific criteria for measuring the quality of management. ARDEC adopted MANTLE as the core of the Center's employee involvement initiative. Using reverse appraisals, ARDEC managers are assessed against the MANTLE model by empowered associates.

In 1995, the quality sword was passed to BG James G. Boddie. He became the driving force behind reinforcing Deming's management principles, performance measurement, and leadership presence. Today, ARDEC's leaders are committed to customer satisfaction, employee involvement, and continuous improvement, an environment that promotes excellence. In 1997, BG Joseph Arbuckle became the ARDEC Commander and Quality Champion. Because of ARDEC's robust approach to continuous improvement, BG Arbuckle was featured on the high profile Total Army Quality "Leading Change" CD-ROM.

The personal leadership of ARDEC's current Commanding General, BG John Geis, its Technical Director, Mr. Michael Fisette, and its senior executive team is the driving force that makes its Vision, Values, and Goals a foundation for all ARDEC does. They have furthered ARDEC's strategic approach while streamlining its systems. Quality and a focus on public responsibility and citizenship, the enhancement of teams, focusing on Baldrige and PAQ criteria, and achievement of real reengineering is ARDEC's strategy for future success in an ever-changing world. In the shift from product to process to total quality, everyone is reaping the benefits—customers, employees, suppliers and, most importantly, the defense of the nation.

ARDEC is frequently recognized for its quality approach. Most recently, ARDEC was declared a winner of the 1999 CSA ACOE Award, 1998 New Jersey Quality Achievement Award (recently renamed the Governor's Award for Performance Excellence), 1996 Presidential Award for Quality, and the 1996 Army Communities of Excellence top award which adoped the Baldrige critera in 1996. ARDEC won the 1995 Quality Improvement Prototype award, was the co-winner of the 1995 Army R&D Organization of the Year and has been ranked in the top 6 of the 20 Army laboratories in this competition eight times in the last 10 years. ARDEC received honorable mention distinction as the Most Improved Organization in the Army Communities of Excellence Program in 1994. It was a DoD nominee for the 1992 Quality Improvement Prototype Award and won the President's Council on Management Improvement Award for Management Excellence in 1991. In addition to the above organization-wide awards, ARDEC teams have received recognition for their results through small team award mechanisms. These awards include: Vice President Gore's Heroes of Reinvention Hammer Award, National Partnership Award, Secretary of Defense Team Excellence Award, Ford Foundation Innovations in American Government (semi-finalist), Federal Executive Board of Northern New Jersey Best Team Award, and the Rochester Institute of Technology/USA TODAY Quality Cup Competition (semi-finalist award).

STATUS OF OUR QUALITY IMPROVEMENT EFFORT

ARDEC's Vision is to "Provide Overwhelming Firepower for Decisive Victory." Its strategic goals are

- ■ Programs - Expand the Customer Base by Securing New Projects.
- ■ Performance - Thrill our Customers by Quickly Meeting their Needs.
- ■ Processes - Provide the Most Cost-Efficient Armament and Munitions Systems.
- ■ People - Attrace, Develop, and Retain a Well-Trained, Well-Equipped and Motivated Workforce.
- ■ Partnerships - Develop Strong Public and Private Partnerships that Increase our Value to the Armed Forces.
- ■ Physical Plant - Develop, Acquire, and Maintain World-Class Facilities and Equipment for Mission Execution.

ARDEC's performance measurement process, the Systems Measurement Review, focuses on results—results based on strong quality criteria and information and analysis. Quality improvement trends and sustained excellent levels of performance are continually demonstrated. ARDEC's System orientation and measurement plans drive the aggregation of performance indicators to better evaluate strategies, communicate requirements, monitor performance, and achieve positive results. Operational, customer-related, and financial results from processes are evaluated together to focus on ARDEC-level improvements and to ensure that results actually reflect better overall performance.

The results ARDEC achieves show three key features of its quality style:

- First, ARDEC's Strategic Planning Process, measurement plans, and results are monitored and evaluated against its Key Business Drivers and are used to generate indicators of contributions toward both strategic and operational results.
- Second, it seeks robust, stable processes, incremental improvements across the board, and innovative leaps in products and processes through benchmarking and competitive comparison.
- Finally, ARDEC nurtures improvement and innovation in its employees. It is aggressive in improving the efficiency of its support systems, and the quality of its products and technology through the use of Concurrent Engineering by Integrated Product Teams.

ARDEC's key customer requirements are maximum effectiveness, improved productivity, reduced life cycle cost, reduced development cycle time, maximum safety for its soldier, and maintaining state-of-the-art technology. Key product effectiveness requirements include accuracy, reliability and lethality. ARDEC's key products are ammunition, guns, fire control, and associated products.

ARDEC ammunition lethality has improved dramatically over the last decade. One critical example is the systematic design improvements during Operation Desert Storm. These Development System improvements resulted in the most effective tank-fired weapon: the -M829A1 ("Silver Bullet"). ARDEC quickly fielded the "Silver Bullet" to meet PM customer requirements in penetrating heavily armored vehicles. Additionally, ARDEC has clear plans to continue to be the best.

ARDEC is focused on improving soldier satisfaction. Implementation of Concurrent Engineering (CE) efforts during the development and testing of fuzes contribute significantly to meeting and satisfying customer requirements. Fuzes are an integral component of ammunition. Successes are attributable to including the Field Artillery School (customers) in the design, test and evaluation phases, where the "soldier" participates in fuze Human Engineering Tests. ARDEC used customer requirements to evolve fuze technology to the point where fuze setting is not only simple and easy, but also exceeds customer requirements by decreasing-fuze setting time by 65% and exceeding the best foreign competitor.

At the same time, ARDEC used computer modeling to reduce testing requirements by approximately 33%, resulting in 60% cost savings.

ARDEC partners with industry and other government agencies to improve quality of items in production. Industry membership is a critical component of the IPTs. The Grenade Machine Gun exemplifies ARDEC's Development and Production System capabilities. The Gun has undergone design changes to enhance its performance and dramatically reduce production costs. ARDEC has improved product reliability exceeding customer requirements and best competitor performance (specifically the thousands of rounds of ammunition fired without stoppages), at reduced costs and improved reliability.

Technical Data Package (TDP) performance is a key Production System requirement from IOC, ACALA, and other acquisition customers. ARDEC must provide quality TDPs in the requested time frames, and TDPs that are immediately usable in acquiring products. ARDEC has dramatically reduced processing time frames and has maintained a level of performance that exceeds the customer delivery requirements with an extremely low defect rate.

The rapid investigation of product malfunction is a Field Support System requirement of ARDEC's IOC customer. ARDEC has improved cycle time to respond to product malfunctions in the field from 10 to 2 days. This was achieved by improvements in communications with the customer, travel management process improvements, and a senior executive commitment.

The Operations and Garrison System Owners are senior executives that lead the effort to achieve the Key Business Driver to "Reduce the cost of doing business." Improved productivity and controlled life cycle costs are key customer requirements. ARDEC continually strives to become more efficient with higher quality. The Productivity Enhancement Group (PEG) studies were concentrated multidisciplined team efforts that provided a road map for a future based structure to absorb downsizing and generate added efficiencies. The result of this strategic systematic planning is the continuing reduction of the overall overhead rate. Via the Financial Restructuring Implementation Team (FRIT), ARDEC used teaming and advanced financial management techniques to overhaul its financial management approach, significantly lowering costs for its customers.

The Human Resources (HR) Plan appendix to the strategic Plan shows how ARDEC's HR plans and initiatives are linked to overall objectives.

ARDEC has HR goals and metrics for short-term and long-term performance. Under the Army IDEAS for Excellence Program, employees are rewarded for ideas that improve processes and/or reduce costs. The HR and Base Operations Sub-System Owners use cross-functional teams to improve processes, establish operational performance goals, and improve overall results.

Quality Deficiency Reports (QDRs) are prepared and sent to ARDEC by soldiers when a product is not performing as expected. A low level of QDRs provides evidence that suppliers are providing the product as designed and without defects.

Reductions in waiver and deviation requests substantiate that ARDEC's suppliers are concentrating on controlling processes and preventing defects. The reduction has led to a substantial decrease in administrative costs.

A key aggregate indicator related to supplier partnering and supplier performance is Value Engineering (VE). ARDEC works closely with its suppliers to reduce the cost to manufacture its designs and shares the savings with the suppliers. The savings to the taxpayer from VE efforts have consistently exceeded the targeted improvement goals. ARDEC has generated increasingly larger savings, which have resulted in $159.6 million over the last 5 years (1994–1998).

CUSTOMER FOCUS

ARDEC's customers are the men and women of the Armed Forces, whose leaders understand that realizing excellence is greatest when the focus is on listening and learning to ensure caring customer service. Former Technical Director, Carmine Spinelli, said it best: *"ARDEC is by its very nature a learning organization. It designs, develops, validates, and changes. More recently it learned to listen."*

ARDEC uses effective feedback systems to obtain knowledge from customers about its products and services. The key method for capturing its diverse customer input is through its six Listening and Learning strategies:

Listen & Learn Strategy 1—Assignment of Customer Advocate Offices. ARDEC recently expanded the role of the PM Advocate to that of an ARDEC Advocate. Several liaisons work with the ARDEC Advocate. Working as a cooperative effort, the ARDEC Advocate and liaisons are responsible for ensuring customer satisfaction for each of its major customer groups. Besides conducting the customer evaluations against the service standards, they attend customer staff meetings and management reviews to oversee and keep current on customer requirements and concerns. The use of focus groups with these key customers help to ascertain future requirements to set better strategies.

Listen & Learn Strategy 2—Integrated Product Teams. Since the IPT is a cross-functional team with members from customer, industry, engineering/technical employees, and support groups, it affords ARDEC another way to get to know its customers. Through this representation, customer needs are expressed on a continual basis. Periodic program reviews with the contractor by IPTs provide an excellent way to obtain continuous customer feedback and influence program development across the spectrum of engineering and support. QFD is a tool integrated into the front end of ARDEC's technical planning process by IPTs. ARDEC and customers continuously communicate throughout the entire product life cycle regarding the requirements and tradeoffs. Through application of QFD and other tools, ARDEC ensures that product designs exceed customer requirements and expectations.

Listen & Learn Strategy 3—Matrix ARDEC employees to the customers. To maintain a close relationship with its key customers, ARDEC representatives are located on-site at customer locations, e.g., PEO/PM, IOC, Army Headquarters, and Battle Labs, to ensure direct and continuous communications. This serves two purposes. First, it makes ARDEC available to the customer for real-time feedback of any issues. Second, it provides a mechanism for quick turn-around of responses to customer concerns.

Listen & Learn Strategy 4—Formal customer review processes. ARDEC uses formal performance review processes to jointly assess its performance with its customers. These include the Technical Director Advisory Council, R&D Reviews, the General Officer Working Group, the Council of Colonels, Milestone Reviews, Production Delivery Reviews and others. ARDEC's Customer Service Principles are

1. Keep your word.
2. Show ownership for resolving problems.
3. Be responsible for timely follow up.
4. Be empowered to make commitments and guarantees.

ARDEC's Customer Contact Service Standards are

1. Reliability: information provided customers is accurate and complete so that the customer can rely on it.
2. Responsiveness: ARDEC responds back to customer requests for information either immediately or (if necessary) not later than 24 hours. Its remediation plans are initiated with the customer within 3 days.
3. Effectiveness: solutions to customer concerns are worked out with the customer and the ARDEC service provider. Remediation plans are not implemented without customer agreement of high success probability.

Listen & Learn Strategy 5—Informal Customer Feedback. ARDEC hosts a variety of customer days. The Military schools, where customer product requirements are generated, are invited to detail their future needs. Engineering and support personnel see and hear customer requirements first hand and the rationale behind those needs along with comments and complaints. This interchange establishes solid customer–ARDEC links with clear understanding of customer needs and technology solutions. Another informal setting for getting to know customers is the Battle Lab breakfasts. A colonel and other ARDEC employees work with the Battle Lab to identify items of interest for discussion as a monthly thematic breakfast.

Listen & Learn Strategy 6—Get information on similar providers. *FOREIGN MANUFACTURERS*—Foreign intelligence exploitation allows ARDEC to benchmark its systems against similar foreign products. This comparison determines how to defend against the product or use the product in its inventory. ARDEC regularly takes foreign rounds and fire them. Or, it fires its weapons against their defenses to determine the degree of vulnerability. Information collected includes the key indicators of ARDEC's own measure of product quality: lethality, range, transportability, and affordability.

> BEST PRACTICE - Leadership- Effective use of cascading communication via chain link teaching.

ARDEC uses video-based training to deploy strategic planning, process management, and customer service principles and standards to the entire workforce.

Leadership links ARDEC's training design approach to its Key Business Drivers. ARDEC calls this process "Training Templates." It began with nine teams identifying courses and curricula. The design of new curricula is a structured process. Mission area teams identify course subjects and content based on core skills identification within the Business Plan. Length of training and target audience are defined. Courses are developed, tested, evaluated, and deployed.

Template Training replaced crude training goals which required minimum hours of training without strategic links to a plan. The HR Development Plan contains the training goals for managers, employees, and teams. The plan provides support for the identification and funding of training for high performance work units. ARDEC focuses curricula development to address the knowledge and skills employees require to meet work objectives; leadership, customer service, team-building, TQM, process management, and quality technical curriculum. The Technical Education and Corporate History (TEACH) program uses senior technical personnel to present seminars.

BEST PRACTICE - Strategic Planning - Effective review process - SMR, including a process which simply but most effectively ensures the continual improvement of key metrics within the organization.

Prior to 1992, ARDEC's macro-level measures were limited to meeting cost, schedule, and short-term productivity. Micro-level measures of performance were disjointed because organizations made selections without focusing on customer satisfaction, unified ARDEC goals, or quality-driven metrics. ARDEC has since developed an information and data process that continually refocuses itself. The ARDEC Vision is clarified by consistent linkage of improvement efforts with customer expectations.

ARDEC has a Measurement Plan that provides guidance on when, how, and what types of measurements will be taken. Process performance is continuously monitored and improved through the use of metrics which are defined in the system measurement.

Measures must be reliable, consistent, relatable to advancing the quality of process or product; and meaningful to customers, suppliers, and performers as an agreed-upon basis for decision-making. Individual performance ratings are tied to this criteria, ensuring that all employees are aware of ARDEC goals and are able to relate how their work contributes to achieving these goals.

ARDEC's emphasis on quality and customer satisfaction leads to measurement systems throughout ARDEC that look at the effectiveness of processes, historical trends, and customer satisfaction. Measurement has a profound impact on the way ARDEC does business. ARDEC indicators are keyed to major life cycle stages. In addition to ARDEC's measurement systems at the macro-level, measurements at all levels of the organization are conducted. Process Action Teams (PAT) and supervisors measure processes using statistical process control and similar techniques to control variability. ARDEC's Systems Measurement Team outputs are used by leadership as its systems level data on performance. Systems measures are automated on a wide area network accessible to organizations, customers and suppliers, system owners, and Executive Council members. Consistency and validity of measures is ensured by customer/supplier input to the performers at the organizational level and executives at the Executive Council level. Direct access to narrative updates allows the tracking of process improvements and future actions. The Systems Measurement Review (SMR) Team provides an assessment of current trends and deviations from goals or projections so that management may make decisions on future plans and programs. To fuel the quality transformation, the analysts who worked on the corporate level information system were trained as quality examiners by the Air Force.

Quarterly, the System Owners review and revise measures linked to ARDEC's strategic goals to assure alignment with customer needs and business priorities. The System Owners are personally responsible for the presentation of the data, analysis, and planning to link customer data, improvements in product and service quality, and improvements in operational performance to financial indicators. Feedback from ARDEC's annual application has resulted in systematic implementation of corrective actions through the personal involvement of senior executives.

The SMR Team compared ARDEC's process to other leaders and adopted guidance from the Air Force Materiel Command (AFMC) Metric Handbook, considered to be the best in its class. AFMC was a 1991 PAQ Recipient. Its Measurement Plan is available on the ARDEC Home Page. The SMR process is defined by the CG Policy and the SMR Desk Guide.

BEST PRACTICE - Information & Analysis - Innovative use of the WWW

ARDEC's use of the World Wide Web (WWW) provides functions (surveys, feedback, document management) and data in many text and graphic formats to customers, managers, and other users. Information made available through such technology is of special advantage in business networks and alliances. Feedback of this information assists teams and System Owners in improving their measurement processes.

ARDEC's Customer Focus Home Page encourages customers to provide feedback in the form of comments, suggestions, and/or complaints. This information is continuously reviewed and problem resolution initiated by the ARDEC Customer Advocate. It is then available to ARDEC service providers and customers. This powerful tool allows current customers, potential customers (which include similar service providers and anyone else on the WWW), and employees to easily view customer rating of products and services. The effect is that service providers are further motivated to provide the best service possible, to quickly resolve any problems and, in general, to develop a "heart" in service providers for satisfying the customer. In this way, customer confidence is effectively retained and restored when required. These strategies are designed for each customer feedback and to show customers ARDEC is proactive in seeking their views. Customers have access to the ARDEC SMR database, the WWW Home Pages, and specific information systems such as TDP Tracker for easy access to data/information. All of these systems incorporate tools for direct and easy customer feedback.

In addition to the Customer Focus Home Page, ARDEC also has placed the results of its frequent Employee Climate surveys on the WWW, as well as its Total Quality Home Page which contains links to its Measurement Plan, and other documents.

ARDEC's Technical Director Michael Fisette stressed the importance of business processes and strategic alliances to ARDEC's approach: "Our responsibility is to build on excellence and execute a vital national security mission. To do that, we must continue to utilize innovative business practices and employ strategic partnerships."

BEST PRACTICE - Process Management - Put in place a deployment process, including a QMB and PATs structured training, an advocacy position (a person who provides training and other support) concepts (e.g., concurrent engineering, benchmarking, and Quality Function Deployment) as they are introduced into the organization.

ARDEC's strength in deploying new concepts throughout its organization is its ability to nurture these initiatives through the use of advocates. ARDEC's advocates are at all levels of the organization, and may be individuals or teams. No matter what the initiative, ARDEC's advocates share knowledge of the issues, well-planned deployment strategies and a passionate commitment to success. ARDEC-wide efforts such as Benchmarking, Integrated Product Teams, MANTLE, Quality Functional Deployment Root Cause Analysis, and Battle Lab coordination have been spurred on through the use of advocates.

Extensive use of QMBs, PATs and Natural Work Groups is one of ARDEC's techniques for channeling employee involvement toward management-selected critical processes fostering a multi-discipline, cross-organizational approach. These teams work toward continuous improvement of ARDEC's major processes. Examples of ARDEC's teams are:

- *The Customer Satisfaction Enhancement Team (CSET)* compared how other similar service organizations did customer surveys and also used customer feedback to formulate an automated customer feedback process. Results included (a) changing the type of quality data collected by having service providers work with their customers to formulate questions based upon expectations; (b) better follow-up with those customers not completing the survey; and (c) an automated quantification and reporting of the ratings. The findings of this team were deployed to other internal continuous customer surveys.

- *Hazardous Materials Management PAT (HMMT):* The objective of this team is to develop a phased Hazardous Materials Management Program (HMMP) for TACOM-ARDEC. Once fully implemented, the HMMP will touch upon operations across the entire arsenal, including tenants and contractors. The value of this improvement will provide Picatinny Arsenal the ability to track hazardous materials and hazardous waste and total asset visibility at a level that has never been available to Picatinny in the past. One measure of ARDEC's success in this initiative is the dollars saved. The 1998 hazardous waste disposal records provide an example of cost savings that can be realized with the implementation of the HSMS and pharmacy. The records show that 76.9% ($224,371.58) of disposal cost was for HAZMAT (unused chemicals). Each step of the process has been monitored to identify methods of delivery, customers, storage, utilization, processes, excess materials/waste and ultimate disposition.

ARDEC is a recognized leader in Concurrent Engineering (CE). Through the use of Integrated Product Teams (IPTs) it systematically researches, designs, develops, tests, produces, and delivers new and/or modified products. IPT characteristics are

- Multifunctional/multidisciplinary team
- Life cycle product responsibility
- Follows ARDEC CE guidelines
- Designs program strategy/plans
- Includes key customers and suppliers
- Team is chartered and empowered

ARDEC's major actions to ensure the full benefits of CE are realized, including:

- Mandatory use of IPTs
- Charters and team training for IPTs
- Entrance and exit criteria applied to each phase of the product life cycle validated by IPTs

ARDEC uses innovative reward and recognition mechanisms to encourage a quality culture and to reinforce behaviors supporting a new team-based climate. For example:

- IPTs that obtain Type Classification for a product receive recognition in multiple forms, e.g., article is put into the ARDEC newspaper "The Voice," an electronic sign at the entrance to the installation announces their accomplishment, a team celebration is conducted, and cash awards are presented to teams.
- The People Enhancing Picatinny (PEP) Rally began in 1987 to celebrate the achievements of the QCs, and is held every year to recognize teams.
- Customer Service Excellence Awards were created to recognize the achievements of service providers.

BEST PRACTICE - Business Results - Effective contractor performance certification program (process for managing suppliers) (CP)

ARDEC partners with many external organizations including suppliers, customers, academia, and industry. ARDEC's partnership efforts with suppliers is evident in the contractor performance certification program (CP)[2]. This program is built around the quality elements of the International Standard Organization (ISO) 9000. Multi-year contracts as well as production options on development contracts are another way ARDEC develops long-term partnerships with its suppliers. ARDEC also partners with suppliers on large development programs using joint teams.

ARDEC's desire to be the dominant, best value supplier to our customers in a competitive downsized DoD is the catalyst for ideas, energy, and partnerships shared between our suppliers, performers, and customers. These ideas, energy, and partnerships have, in turn, driven ARDEC to a pervasive search for quality in product and processes.

ARDEC utilizes a preventive approach through its robust safety, security, emergency services, and environmental programs. ARDEC created a centralized risk management Directorate of Public Safety and Environmental Affairs to ensure an integrated and proactive approach to fire protection and emergency services, safety, security, environmental compliance, and environmental restoration. Its network of organizational safety coordinators and organizational environmental coordinators augment and internalize safety and environmental concerns into all practices. A comprehensive set of measures monitors performance. Analysis of safety data is performed by safety and environmental specialists who report bi-weekly to the Command group. Environmental progress is reported quarterly to the commander as part of the Systems Management Review process.

REFERENCES

1. Joiner, B.L. *Fourth Generation Management: The New Business Consciousness*, McGraw-Hill, 1994, pp. 43–51.
2. Brown, M.G. *Baldrige Award Winning Quality*, 4th ed., Quality Resources and ASQC Quality Press, 1994, p. 350.
3. Deming, W.E. *The New Economics for Industry, Government, and Education*. Massachusetts Institute of Technology Center for Advanced Engineering Study, Cambridge, MA, 1993, pp. 99–109.
4. Helton, B.R. The Baldie Play. *Quality Progress* (February 1995): 43–45.
5. Quinn, J.B. *Intelligent Enterprise: A Knowledge and Service Based Paradigm for Industry*. The Free Press, New York, 1992, pp. 3–5.
6. Clavell, J. (Transl.). *Sun Tzu, Art of War*. Dell, New York, 1983, pp. 20–22, 29.

INDEX

A

AARS (Associate Administrator Review Status), KSC, 305

Action plan teams
action sequence for results, 25–29
ARDEC, 351
BI Performance Services, 181
ESQA feedback implementation process, 93
steering committee, TVA F&HP, 279–281
UWMT, 215, 237–240

Act phase, PDKA theory
information into knowledge, 8
PQAP feedback, 84–85
TVA F&HP implementation of, 275–286

ARDEC (U.S. Army Armament Research, Development and Engineering Center)
action plan teams, 351
background, 312, 337–338
benchmarking studies, 351
business overview, 338–342
business results, 357
customer feedback, 350–352
customer groups, 342–343
customer requirements, 348–350
employee motivation, 357
executive buy-in requirement, 313, 338
information consolidation techniques, 314, 354–358
mission, 341, 343–344
operational results, 348–349
organizational structure, 341–342
organizing framework, 315, 338
periodic reviews, 314, 340
problem-solving process, 315
products and services, 342
quality pursuit history, 344–347

strategic goals, 348
supplier relationships, 343–344
training strategy, 340–341, 353

Army Communities of Excellence Award, 318, 338

Associate Administrator Review Status (AARS), KSC, 305

B

Baldrige-style categories, 25, 38–39

Benchmarking strategies
ARDEC, 352
FSS NECR, 262–263
internal, 72
KSC, 246, 301–303
Pal's Sudden Service, 129–130, 141–144
Royal Mail, 223–224, 227
Ulster Carpet Mills, 168

BI Performance Services
action plan teams, 181
background, 179–180, 187–188
customer focus, 188–189
decision-making criteria, 183
information consolidation techniques, 181–182, 196–197
integrated solutions examples, 189–190
mission, 188
organizing framework, 182–183, 192–197
periodic reviews, 184–185
scope of services, 189
strategic measurement/analysis model, 192
strengths and weaknesses identified, 183
success factors, 190–191
task assignment, 184, 200
TCSI, 193–199